Emotional Intelligence and Neuro-linguistic Programming

Manufacturing Design and Technology Series

Series Editor:
J. Paulo Davim

This series will publish high quality references and advanced textbooks in the broad area of manufacturing design and technology, with a special focus on sustainability in manufacturing. Books in the series should find a balance between academic research and industrial application. This series targets academics and practicing engineers working on topics in materials science, mechanical engineering, industrial engineering, systems engineering, and environmental engineering as related to manufacturing systems, as well as professions in manufacturing design.

Drills: Science and Technology of Advanced Operations
Viktor P. Astakhov

Technological Challenges and Management: Matching Human and Business Needs
Edited by Carolina Machado and J. Paulo Davim

Advanced Machining Processes: Innovative Modeling Techniques
Edited by Angelos P. Markopoulos and J. Paulo Davim

Management and Technological Challenges in the Digital Age
Edited by Pedro Novo Melo and Carolina Machado

Machining of Light Alloys: Aluminum, Titanium, and Magnesium
Edited by Diego Carou and J. Paulo Davim

Additive Manufacturing: Applications and Innovations
Edited by Rupinder Singh and J. Paulo Davim

For more information about this series, please visit: https://www.crcpress.com/Manufacturing-Design-and-Technology/book-series/CRCMANDESTEC

Emotional Intelligence and Neuro-linguistic Programming
New Insights for Managers and Engineers

Edited by
Carolina Machado and J. Paulo Davim

CRC Press
Taylor & Francis Group
Boca Raton London New York

CRC Press is an imprint of the
Taylor & Francis Group, an **informa** business

CRC Press
Taylor & Francis Group
6000 Broken Sound Parkway NW, Suite 300
Boca Raton, FL 33487-2742

First issued in paperback 2020

© 2019 by Taylor & Francis Group, LLC
CRC Press is an imprint of Taylor & Francis Group, an Informa business

No claim to original U.S. Government works

ISBN-13: 978-1-138-04974-1 (hbk)
ISBN-13: 978-0-367-77954-2 (pbk)

Library of Congress Cataloging-in-Publication Data

Names: Machado, Carolina, 1965- editor. | Davim, J. Paulo, editor.
Title: Emotional intelligence and neuro-linguistic programming: new insights for managers and engineers / edited by Carolina Machado and J. Paulo Davim.
Description: Boca Raton, FL: CRC Press, 2019. |
Series: Manufacturing design and technology series | Includes bibliographical references and index.
Identifiers: LCCN 2019008217 | ISBN 9781138049741 (hardback: acid-free paper) | ISBN 9781315168906 (ebook)
Subjects: LCSH: Psychology, Industrial. | Neurolinguistic programming. | Emotional intelligence. | Management. | Organizational behavior.
Classification: LCC HF5548.8 .E49195 2019 | DDC 152.402/462—dc23
LC record available at https://lccn.loc.gov/2019008217

Visit the Taylor & Francis Web site at
http://www.taylorandfrancis.com

and the CRC Press Web site at
http://www.crcpress.com

Contents

Contents

Preface

Emotional Intelligence (EI) aims to describe the ability to recognise and evaluate our own feelings and the others' feelings, as well as the ability to deal with them. As an approach to communication and personal development, Neurolinguistic Programming (NLP) contributes to a better understanding of our internal functioning, to identify our mental models in order to evaluate, reflect and redesign them. Joined, they were a great challenge to managers and engineers, as they have a deep impact in the business world, particularly in the areas of leadership and employee development. NLP is an extraordinary technology, since it allows us to unconsciously envision new and improved choices, even before we can consciously select our usual choice. This is the real secret of flexibility; in other words, of EI.

Today, all of us are continuously facing new challenges in our daily and work life. In the workplace, we face many and diverse challenges and problems relating to poor communication, unclear objectives, pressure, making decisions, meeting deadlines, etc. More than technical skills, many employees don't obtain necessary and potential performance levels due to a lack of self-confidence, anxiety and demotivation. At this level, EI assumes a relevant role as a means to work with our emotions in an intelligent way. Developing employees' emotional intelligence is a powerful contribution to self-awareness, expressing emotions, making decisions, interpersonal relationships and dealing with stress. But in order to obtain better results, it is important that we know more about ourselves as a way of increasing our attitudinal and behavioural choices. At this level, NLP appears as relevant methodology that helps to achieve the desired results. With NLP, it is possible to teach employees to manage their thoughts and emotions, becoming the leader of themselves and of their relationships with others, thereby increasing their performance. Together, EI and NLP are important methodologies that managers and engineers can use in their daily routines in their organisation. Conscious of its importance to organisations nowadays, this book looks to highlight the contributions that different academics, researchers and practitioners are giving to these matters in a diverse range of organisations all over the world.

Aiming to promote research related to new trends that open up a new field of research in the management and engineering areas, the present book, divided into five chapters, covers: in chapter one, *"The relationship between emotional intelligence and job performance: an evidence-based literature review with implications for scientists and engineers"*; in chapter two, *"Emotional intelligence: telling the history and discovering the models"*; in chapter three, *"How Japanese managers use NLP in their daily work"*; in chapter four, *"Neurolinguistic programming for managers and engineers as evidence-based practitioners"*; and, finally, in chapter five, *"The evolution of cultural intelligence (CQ) and its impact on individuals and organisations"*.

Focusing on the latest and most recent research findings occurring in this field and looking to share knowledge and insights on an international scale, this book, entitled *Emotional Intelligence and Neurolinguistic Programming: New challenges to Managers and Engineers,* can be used as a research book for undergraduate engineering and management courses or as a subject on management and engineering at the postgraduate level. Also, this book can serve as a useful reference for academics, researchers, managers, engineers and other professionals in related fields.

The Editors acknowledge their gratitude to CRC Press/Taylor & Francis Group for this opportunity and for their professional support. Finally, we would like to thank all chapter authors for their interest and availability to work on this project.

Carolina Machado
Braga, PORTUGAL

J. Paulo Davim
Aveiro, PORTUGAL

Editors

Carolina Machado received her Ph.D. degree in Management Sciences (Organisational and Policies Management area/Human Resources Management) from the University of Minho in 1999, Master's degree in Management (Strategic Human Resource Management) from Technical University of Lisbon in 1994 and Degree in Business Administration from University of Minho in 1989. Teaching Human Resources Management subjects since 1989 at University of Minho, she is since 2004 Associate Professor, with experience and research interest areas in the field of Human Resource Management, International Human Resource Management, Human Resource Management in SMEs, Training and Development, Emotional Intelligence, Management Change, Knowledge Management and Management/HRM in the Digital Age. She is Head of the Department of Management and Head of the Human Resources Management Work Group at University of Minho, as well as Chief Editor of the International Journal of Applied Management Sciences and Engineering (IJAMSE), Guest Editor of journals, books Editor and book Series Editor, as well as reviewer in different international prestigious journals. In addition, she has published both as editor/co-editor and as author/co-author several books, book chapters and articles in journals and conferences.

J. Paulo Davim received his Ph.D. degree in Mechanical Engineering in 1997, M.Sc. degree in Mechanical Engineering (materials and manufacturing processes) in 1991, Mechanical Engineering degree (5 years) in 1986, from the University of Porto (FEUP), the Aggregate title (Full Habilitation) from the University of Coimbra in 2005 and the D.Sc. from London Metropolitan University in 2013. He is also Eur Ing by FEANI-Brussels and Fellow (FIET) by IET-London. Senior Chartered Engineer by the Portuguese Institution of Engineers with an MBA and Specialist title in Engineering and Industrial Management. Currently, he is Professor at the Department of Mechanical Engineering of the University of Aveiro, Portugal. He has more than 30 years of teaching and research experience in Manufacturing, Materials, Mechanical and Industrial Engineering, with special emphasis in Machining & Tribology. He also has interest in Management, Engineering Education and Higher Education for Sustainability. He has guided large numbers of postdoc, Ph.D. and master's students as well as coordinated and participated in several financed research projects. He has received several scientific awards. He has worked as evaluator of projects for ERC-European Research Council and other international research agencies as well as examiner of Ph.D. thesis for many universities in different countries. He is the Editor-in-Chief of several international journals, Guest Editor of journals, books Editor, book Series Editor and Scientific Advisory for many international journals and conferences. Presently, he is an Editorial Board member of 30 international journals and acts as reviewer for more than 100 prestigious Web of Science journals. In addition, he has also published as editor and co-editor of more than 100 books and as author and co-author of more than 10 books, 80 book chapters and 400 articles in journals and conferences (more than 250 articles in journals indexed in Web of Science core collection/h-index 50+/7500+ citations, SCOPUS/h-index 56+/10,000+ citations, Google Scholar/h-index 71+/16,000+).

Contributors

William Van Gordon
Department of Online
    Learning
University of Derby
Derby, United Kingdom

**Slawomir Jarmuz Stawomir
Jarmuz**
Moderator Ltd
Wroctaw, Poland

Yasuhiro Kotera
Department of Online
    Learning
University of Derby
Derby, United Kingdom

Mark L. Lengnick-Hall
College of Business
University of Texas at San
    Antonio
San Antonio, Texas

Carolina Machado
Department of Management
University of Minho
Braga, Portugal

Cláudia Sofia Fanha Moura
Department of Management
University of Minho
Braga, Portugal

Ana Lúcia Rodrigues
Department of Management
University of Minho
Braga, Portugal

Christopher B. Stone
Barton School of Business
Wichita State University
Wichita, Kansas

Tomasz Witkowski
Polish Skeptics Club
Wrocław, Poland

chapter one

The relationship between Emotional Intelligence and job performance

An evidence-based literature review with implications for scientists and engineers

Mark L. Lengnick-Hall
University of Texas at San Antonio

Christopher B. Stone
Wichita State University

Contents

Introduction

Emotional intelligence (EI)—also known as emotional competence—has been a focus of research interest for several decades. At its most general description, EI has been defined as the "ability to monitor one's own and others' feelings and emotions, to discriminate among them, and to use this information to guide one's thinking and actions" (Salovey & Mayer, 1990, p. 189). Joseph and Newman (2010) report that EI has stimulated interest among human resource practitioners who have embraced the concept and used it in personnel hiring and training. They provide evidence of this interest by identifying 57 consulting firms dedicated primarily to EI, 90 organisations that specialise in training or assessment of EI, 30 EI certification programmes, and 5 EI "universities"—and this was in 2008. We can only imagine how this EI "industry" has grown since then.

There continues to be debate among those who consider EI a major organisational tool versus those who question its utility and scientific foundation. The purpose of this chapter is to provide an evidence-based review of existing high-quality research and scientific knowledge regarding the relationship between EI and job performance. Additionally, we provide implications for the role of EI in the job performance of scientists and engineers.

We follow the principles of evidence-based management practice as outlined by Barends, Rousseau, and Briner (2017): (1) translating a practical issue into an answerable question, (2) systematically searching for and retrieving evidence, (3) critically judging the evidence, (4) pulling together the evidence, (5) incorporating the evidence into the decision-making process, and then (6) evaluating the outcome of the decision taken. Our focus in this chapter is on steps (1)—(4). We leave steps (5) and (6) to decision-makers in organisations.

In this chapter, we ask specific questions about the relationship between EI and job performance, systematically search for and retrieve a particularly high-quality source of evidence (published and peer-reviewed articles), and critically examine the quality and implications of the findings. We also examine and assess the pattern and extent of such evidence.

Emotional intelligence

Part of the controversy regarding EI hinges on construct clarity. Two alternative approaches have been taken in defining EI. In one approach—the ability-based approach—EI is viewed as a type of intelligence or aptitude that overlaps logically with cognitive ability. This is the approach taken by Salovey and Mayer (1990) who define EI as a set of abilities that includes four branches or dimensions, including (1) accurately perceiving emotions in one's self and others (*emotional perception*); (2) using emotions to facilitate thinking (*emotional facilitation*); (3) understanding emotions, emotional language, and the signals conveyed by emotions (*emotional understanding*); and (4) managing emotions so as to attain specific goals (*emotional regulation*). In contrast, mixed EI models view EI as a combination of intellect and various measures of personality and affect (Joseph and Newman, 2010). To further complicate the issue, many believe ability-based EI to not only have a stronger theoretical foundation but also be a weaker predictor of job performance, whereas mixed EI is believed to have a weaker theoretical foundation but is believed to be a stronger predictor of job performance.

In this chapter, we seek answers to the following questions:

1. Is there a relationship between EI and job performance? If so, how strong is it?
2. Does EI incrementally predict job performance when measures of personality and cognitive intelligence are included as predictors?
3. What mediates the relationship between EI and job performance?
4. What moderates the relationship between EI and job performance?

Methodology

We searched three major multidisciplinary databases: (1) ABI/INFORM Collection, (2) Business Source Complete, and (3) PsycINFO. ABI/INFORM Collection provides access to three electronic resources which can be searched simultaneously: (1) ABI/INFORM Dateline, (2) ABI/INFORM Global, and (3) ABI/INFORM Trade & Industry. ABI/INFORM Global indexes and abstracts over 1,800 business periodicals. ABI/INFORM Trade & Industry offers more than 750 specific trade and industry publications, while ABI/INFORM Dateline provides access to 170 local, city, state, and regional business publications. Business Source Complete provides full-text coverage for over 9,400 general and scholarly business journals and other sources in marketing, management, economics, finance, accounting, and international business. PsycINFO covers professional and academic literature in psychology and related disciplines, including medicine, psychiatry, sociology, education, and linguistics. It indexes and abstracts articles in more than 1,900 journals worldwide.

Our procedure across these databases involved searching for article titles and abstracts for the terms "emotional intelligence" OR "emotional competence" AND "job performance" or "work performance." Our goal was to narrow our results only to articles examining the relationship between EI and job performance. We searched exclusively for articles in which our key search terms (emotional intelligence, emotional competence, job performance, and work performance) were included in the title or the abstract to ensure that we only identified articles that were specifically about our topic of interest and excluded articles which did not examine the relationship between EI and job performance.

Our search of the three publication databases resulted in 303 peer-reviewed articles. We then eliminated articles (as done by

Marler & Boudreau, 2017) in journals that were not on the Journal Quality List (JQL), 63rd edition, July 29, 2018 (Harzing, 2018). Next, we eliminated duplicate articles from our list. For the remaining 27 articles in scholarly peer-reviewed journals, we extracted the following information: (1) author and year of publication, (2) sector/population, (3) design and sample size, (4) main findings, (5) effect size, (6) moderators/mediators, (7) limitations, and (8) design level (using criteria by Barends, Rousseau, & Briner, 2017). See Table 1.1.

Critical evaluation of evidence

In order to answer our research questions, we reviewed the information collected in Table 1.1, as well as examining the retained articles in more detail. We looked for patterns across the studies including: (1) Sector/population—Were the sectors/populations more homogeneous or more heterogeneous in terms of types of participants and countries where participants resided? (2) Design and sample size—Were stronger or weaker research designs used by researchers? Were studies able to provide stronger or weaker evidence for causality when assessing the relationship between EI and job performance? Were samples large or small? (3) Main findings—Was the evidence consistent? (i.e., pointing to identical or similar conclusions?); Was the evidence contested? (one or more study/studies directly refutes or contests the findings of other studies, raising questions about the trustworthiness of the purported effect?); Was the evidence mixed? (studies based on a variety of different designs or methods, applied in a range of contexts, have produced results that suggest underlying difference in the nature of the effects observed or important differences across studies that are not yet well understood?).

Results

Our discussion of results is organised around the key research questions posed at the beginning of the chapter.

Is there a relationship between EI and job performance? If so, how strong is it?

First, how is job performance typically operationalised in EI research? Task performance is the core substantive duties that are formally recognised as part of the job; but in addition to task performance, organisational citizenship behaviours (OCB) or contextual performance concerns activities that contribute to the achievement of the objectives of an organisation, but are not formally recognised (Cote & Miners, 2006). In EI research, objective task performance is rarely measured. More typically, EI research relies

Table 1.1 Categorisation of peer-reviewed journal publications

Author (year)	Sector/ Population	Design + Sample size	Main findings	Effect size	Moderators/ Mediators	Limitations	Design level
Blickle et al. (2009)	Working adults in Bonn, Germany	210 working adults (71.4% female), longitudinal, three waves, questionnaires to participants and their supervisors, peers, and subordinates, random assignment for third wave (General Mental Ability—GMA)	Emotional intelligence (EI) explained additional variance in overall job performance ratings beyond GMA and personality traits	Assessments of overall job performance by supervisors, peers, and others correlated significantly with TEMINT, which is a measure of emotional reasoning skills ($r = -0.24$). Emotional reasoning skills explained a significant additional proportion of variance in overall job performance beyond GMA and personality (i.e., the Big Five). Standardised beta weight was ($\beta = -0.21$)	NA	Only modest correlations; no parallel-test reliabilities of TEMINT available; archival performance data from HR not available for comparison	B

(Continued)

Table 1.1 (Continued) Categorisation of peer-reviewed journal publications

Author (year)	Sector/ Population	Design + Sample size	Main findings	Effect size	Moderators/ Mediators	Limitations	Design level
Bozionelos and Singh (2017)	Full-time expatriate employees in the United Arab Emirates (UAE)	188 (48 women and 140 men); subordinate-line manager dyads responded to one-time survey	EI & EI facets had a U-shaped relationship with job performance	EI & task performance ($r = 0.19$); EI & OCB-H ($r = 0.19$); EI & OCB-V ($r = 0.18$)	NA	Cross-sectional design; limited generalisability	D
Carmeli (2003)	CFOs in local government authorities in Israel	98 (11 women); cross-sectional (questionnaire)	EI positively related to job performance	EI to job performance ($\beta = 0.32$)	EI weakened the negative effect of work–family conflict on career commitment	Did not include non-managerial employees; self-report data; new measure of EI	D

(Continued)

Table 1.1 (Continued) Categorisation of peer-reviewed journal publications

Author (year)	Sector/ Population	Design + Sample size	Main findings	Effect size	Moderators/ Mediators	Limitations	Design level
Christiansen, Janovics, and Siers (2010)	Employed students at a U.S. university	175; one-time survey	Performance-based EI correlated more strongly with job performance than did that of the self-report but provided little incremental validity beyond cognitive ability and conscientious-ness	Performance-based EI factor correlated more strongly with job performance ($\rho = 0.24$) than did that of the self-report ($\rho = 0.05$)	NA	College students (although working); limited to lower-level jobs; limited generalis-ability	D

(Continued)

Table 1.1 (Continued) Categorisation of peer-reviewed journal publications

Author (year)	Sector/ Population	Design + Sample size	Main findings	Effect size	Moderators/ Mediators	Limitations	Design level
Clarke and Mahadi (2017)	Employees in large insurance organisation in Malaysia	203 matched leader–follower dyads; (54% of leaders male); survey with independent and dependent variables collected at two different times	Leader EI had both direct and indirect effects on follower job performance	Leader EI & follower job performance ($\beta = 0.07$)	NA	Cross-sectional design; limited generalis-ability; potential common method bias; potential social desirability bias; subjective measure of job performance	D

(Continued)

Table 1.1 (Continued) Categorisation of peer-reviewed journal publications

Author (year)	Sector/ Population	Design + Sample size	Main findings	Effect size	Moderators/ Mediators	Limitations	Design level
Cote and Miners (2006)	Employees of a large public university	175 full-time employees (67% female); one-time survey	The association between EI and task performance will become more positive as cognitive intelligence decreases. The interaction between EI and cognitive intelligence predicts OCBO	EI & job performance ($r = 0.32$); EI & OCBO ($r = 0.65$); EI & OCBI ($r = 0.60$)	Cognitive intelligence	Cross-sectional design; limited generalis-ability; subjective performance ratings	D
Dulewicz, Higgs, and Slaski (2003)	Middle-managers of a large retail organisa-tion	59 middle-managers (39% women); survey	EIQ positively related to job performance	EIQ to job perfor-mance ($r = 0.321$)	NA	Small sample size; one single private sector retail organisation	D

(Continued)

Table 1.1 (Continued) Categorisation of peer-reviewed journal publications

Author (year)	Sector/ Population	Design + Sample size	Main findings	Effect size	Moderators/ Mediators	Limitations	Design level
Farh, Seo, and Tesluk (2012)	Current employees in MBA pro-gramme in Mid-Atlantic U.S.	346 full-time professionals (70% male); two surveys—one completed by employees (EI), another completed by supervisors (ratings of teamwork and job performance)	EI–performance relationship is contingent on an MWD (manage-rial work demands) job context	Cannot extract simple effect size from mediated model	An MWD job context moderates the positive relationship between EI and teamwork effectiveness, such that the relationship is stronger for employees working in higher rather than lower MWD job contexts (supported). The moderat-ing effect of an MWD job context on the relationship between EI	Cross-sectional design; limited generalis-ability	D

(Continued)

Table 1.1 (*Continued*) Categorisation of peer-reviewed journal publications

Author (year)	Sector/ Population	Design + Sample size	Main findings	Effect size	Moderators/ Mediators	Limitations	Design level
					and job performance is mediated by teamwork effectiveness (supported)		
Greenidge, Devonish, and Alleyne (2014)	Employees in organisations in the Caribbean	222 employees (45.5% male, 54.5% female) across five organisations from the manufacturing, financial, and services private sector industries, and the public sector in the English-speaking Caribbean; responded to one-time survey	Job satisfaction partially mediated the relationship between (a) the ability-based EI dimension, regulation of emotion, and contextual performance, and the ability-based EI dimension, use of emotion and contextual performance	EI dimensions & job performance: self-emotion appraisal ($r = 0.20$), others' emotion appraisal ($r = 0.18$), use of emotion ($r = 0.22$), & regulation of emotion ($r = 0.25$)	Job satisfaction partially mediated the relationship between (a) the ability-based EI dimension, regulation of emotion, and contextual performance, and the ability-based EI dimension, use of emotion and contextual performance	Potential bias in using self-selected co-workers to rate performance; cross-sectional study; model may be under-specified	D

(Continued)

Table 1.1 (Continued) Categorisation of peer-reviewed journal publications

Author (year)	Sector/ Population	Design + Sample size	Main findings	Effect size	Moderators/ Mediators	Limitations	Design level
Huang, Chan, Lam, and Nan (2010)	Leader-member dyads from call centre of a telecommunications services company in China	493 leader-member dyads; one-time survey	Burnout mediated the link between use of emotion and work performance. LMX was associated with burnout and work performance more strongly for service workers with lower levels of self-emotion appraisal. The link between LMX and work performance was stronger for service workers with higher levels of use of emotion	Cannot extract simple effect size from mediated model	Burnout as a mediator; self-emotion appraisal as a moderator	Self-reported measure of LMX & burnout (potential common method variance); no evidence of direction of relationships; limited generalisability	D

(Continued)

Table 1.1 (Continued) Categorisation of peer-reviewed journal publications

Author (year)	Sector/ Population	Design + Sample size	Main findings	Effect size	Moderators/ Mediators	Limitations	Design level
Joseph and Newman (2010)	21 published meta-analytic correlations plus 66 original meta-analyses	Meta-analysis	EI positively predicts performance for high-emotional labour jobs and negatively predicts performance for low-emotional labour jobs	EI facet of emotion regulation & job performance ($\beta = 0.18$)	NA	Paucity of studies using actual job performance as a criterion	B
Joseph, Jin, Newman, and O'Boyle (2015)	20 published meta-analytic correlations plus 16 original meta-analyses	Meta-analysis	An updated estimate of the meta-analytic correlation between mixed EI and supervisor-rated job performance is ($p = 0.29$)	Mixed EI and job performance ($p = 0.29$)	NA	Did not control for all relevant personality traits; only considered average effects across jobs	C

(Continued)

Table 1.1 (Continued) Categorisation of peer-reviewed journal publications

Author (year)	Sector/ Population	Design + Sample size	Main findings	Effect size	Moderators/ Mediators	Limitations	Design level
Kim and Liu (2017)	Hong Kong Chinese newcomer employees	137 newcomers (47% female); two-wave survey with independent and dependent variables collected at two different times	Emotional competence significantly moderated the relationship between taking charge and job performance, such that taking charge was positively related to job performance only when newcomers' emotional competence was high	Taking charge moderated by EI related to job performance ($\beta = 0.21$)	Emotional competence moderates the relationship between taking charge and employees' job performance, such that the relationship is positive only when emotional competence is high	Limited generalisability; potential common method bias; subjective job performance measure; failed to control for some variables known to influence socialisation	D

(Continued)

Table 1.1 (Continued) Categorisation of peer-reviewed journal publications

Author (year)	Sector/ Population	Design + Sample size	Main findings	Effect size	Moderators/ Mediators	Limitations	Design level
Kluemper, DeGroot, and Choi (2013)	Current employees in MBA pro- gramme in U.S. South; individuals hired as treatment staff members at a large residential treatment centre in U.S. Midwest	220 employees (55% female)— MBA pro- gramme, one-time survey; 100 employees (49% female)— residential treatment centre, predic- tive validation study	EI consistently demonstrates incremental validity and is the strongest relative predictor of task perfor- mance and individually directed OCB	MBA sample: EI and task performance ($r = 0.30$), EI and OCB-I ($r = 0.25$), EI and OCB-O ($r = 0.24$); residential treatment sample: EI and task performance ($r = 0.27$), EI and OCB-I ($r = 0.23$), EI and OCB-O ($r = 0.21$)	NA	Small sample sizes; participants were relatively young adults; many participants worked part-time	C

(Continued)

Table 1.1 (Continued) Categorisation of peer-reviewed journal publications

Author (year)	Sector/ Population	Design + Sample size	Main findings	Effect size	Moderators/ Mediators	Limitations	Design level
Law, Wong, and Song (2004)	Employees of a cigarette factory in Anhui province	Cross-sectional (questionnaire); 165 response sets (supervisors, employees, peers)	Peer ratings of EI were found to be significant predictors of job performance ratings provided by supervisors	Peer rating of EI positively related to task performance ($\beta = 0.42$)	NA	Self-report; data collected in Hong Kong and People's Republic of China	D
Law, Wong, Huang, and Li (2008)	Research & development scientists in China	102 employees (78% male); one-time survey	EI is a significant predictor of job performance beyond the effect of the GMA battery on performance ($\Delta R^2 = 0.10$)	For two dimensions of WLEIS measure, other's emotional appraisal & job performance ($r = 0.26$); emotional regulation & job performance ($r = 0.20$)	NA	Cross-sectional study; limited generalisability	D

(Continued)

Table 1.1 (Continued) Categorisation of peer-reviewed journal publications

Author (year)	Sector/Population	Design + Sample size	Main findings	Effect size	Moderators/Mediators	Limitations	Design level
Lindebaum (2013)	Two public sector organisations in the UK	67 team members (20 male, 47 female), 70 line managers completed one-time survey	Preliminary evidence that emotional intelligence moderates the relationship between mental health and job performance (i.e., job dedication as one dimension of contextual performance)	Somatic anxiety & EI used as predictor variables for contextual performance dimension of job dedication ($R^2 = 0.33$)	No support for hypothesis that EI moderates the relationship between mental health & task performance	Small sample size—low power; limited generalisability	D

(Continued)

Table 1.1 (Continued) Categorisation of peer-reviewed journal publications

Author (year)	Sector/ Population	Design + Sample size	Main findings	Effect size	Moderators/ Mediators	Limitations	Design level
O'Boyle, Humphrey, Pollack, Hawver, and Story, 2011	Classified EI studies into three streams: (1) ability-based models that use objective test items, (2) self-report or peer-report measures based on the four branch model of EI, and (3) "mixed models" of emotional competencies	Meta-analysis	Overall relation between EI and job performance is ($r = 0.28$). The three streams have corrected correlations ranging from 0.24 to 0.30 with job performance	Overall relation between EI and job performance is ($r = 0.28$); the three streams have corrected correlations ranging from 0.24 to 0.30 with job performance	NA	Limited by moderators reported in literature; job performance only focused on task performance; range restriction in job performance	C

(Continued)

Table 1.1 (Continued) Categorisation of peer-reviewed journal publications

Author (year)	Sector/ Population	Design + Sample size	Main findings	Effect size	Moderators/ Mediators	Limitations	Design level
Ono, Sachau, Deal, Englert, and Taylor (2011)	Federal law enforcement agents in the U.S.	131 federal law enforcement agents across the United States; surveys at two times; validation study	EI predicts some aspects of investigator performance better than the five personality dimensions and cognitive ability	EI & job performance ($\beta = 0.36$)	NA	Small sample size; used mixed-model EI measure	B
Pekaar, van der Linden, Bakker, and Born (2017)	Sample of Dutch divorce lawyers and salespersons	57 divorce lawyers (94.1% female), 62 salespersons (62.3% male); diary study	EI dimension of others-emotion appraisal contributed more to subjective and objective job performance than other EI dimensions	Other-focused EI & subjective job performance ($\beta = 0.50$); other-focused EI & subjective job performance ($\gamma = 1.593$)	EI dimensions interacted with regard to job performance, such that appraising the emotions of one person was more effective than of two persons (other and self), and appraising others' emotions was more effective	Didn't control for cognitive intelligence or personality; some reliability issues; potential social desirability	D

(Continued)

Table 1.1 (Continued) Categorisation of peer-reviewed journal publications

Author (year)	Sector/ Population	Design + Sample size	Main findings	Effect size	Moderators/ Mediators	Limitations	Design level
					when one's own emotions were also used or regulated		
Sastre Castillo and Danvila Del Valle (2017)	Employees in low-skilled back-office positions from a service company in Spain	405 employees responded to one-time survey	EI was related to job performance. Predictive power was increased when EI & affective commitment considered simultaneously	EI & job performance ($\beta = 0.14$)	NA	Cross-sectional design; limited generalis-ability; potential common method bias	D
Semadar, Robins, and Ferris (2006)	Large, Australian-based, motor manufac-turing company	136 managers (79% male); one-time survey	EI exhibited a significant bivariate association with job performance, but it was not a significant predictor of performance in	EI & job performance ($r = 0.25$)	NA	Limited scope in examining scope of managerial effective-ness; one-item measure to	D

(Continued)

Table 1.1 (Continued) Categorisation of peer-reviewed journal publications

Author (year)	Sector/ Population	Design + Sample size	Main findings	Effect size	Moderators/ Mediators	Limitations	Design level
			the context of the other social effectiveness predictors			assess job performance; generic managerial role	
Shih and Susanto (2010)	Public organisation in Indonesia	228 (65.4% male), one-time survey	Integrating conflict management style partially mediates the relationship between EI and job performance	Cannot extract simple effect size from mediated model	Mediator— integrating conflict management style	Cross-sectional study; limited generalisability; self-reported measures (potential common method variance, positive affectivity, & social desirability)	D

(Continued)

Table 1.1 (Continued) Categorisation of peer-reviewed journal publications

Author (year)	Sector/ Population	Design + Sample size	Main findings	Effect size	Moderators/ Mediators	Limitations	Design level
Sy, Tram, and O'Hara (2006)	Employees of a restaurant franchise	187 food service workers (61% female) and their 62 managers (66% female); one-time survey	Employees' EI positively predicts job performance after controlling for the Big Five personality factors ($R^2 = 0.03$)	Employee EI & job performance ($r = 0.28$); manager's EI and job performance ($r = 0.18$)	NA	EI measured by self-report; subjective measure of performance, younger participants	D
Vidyarthi, Anand, and Liden (2014)	Assembly-line workers and their supervisors in a large multinational manufacturing organisation in India	88 managers, 391 full-time employees, one-time survey	Relationships between leaders' emotion perceptions and employees' job performance was strengthened by task interdependence and attenuated by power distance	Leader's emotion perceptions related to employees' job performance ($\beta = 0.58$)	Relationships between leaders' emotion perceptions and employees' job performance was strengthened by task interdependence and attenuated by power distance	Cross-sectional design; control variables not comprehensive	D

(Continued)

Table 1.1 (Continued) Categorisation of peer-reviewed journal publications

Author (year)	Sector/ Population	Design + Sample size	Main findings	Effect size	Moderators/ Mediators	Limitations	Design level
Weinzimmer, Baumann, Gullifor, and Koubova (2017)	Individuals recruited through Facebook	Web survey; 233 participants (54% women)	EI indirectly related to job performance through the mediator of work–family balance	EI to performance rating ($r = 0.20$)	Work–family balance mediated the relationship between EI and job performance	Common method bias; single item measure of job performance; type of EI measurement (ability-based, self-report)	D

(Continued)

Emotional Intelligence and Neuro-linguistic Programming

Table 1.1 (Continued) Categorisation of peer-reviewed journal publications

Author (year)	Sector/ Population	Design + Sample size	Main findings	Effect size	Moderators/ Mediators	Limitations	Design level
Wilderom, Hur, Wiersma, van den Berg, and Lee (2015)	Managers & employees of a large retail electronics chain in South Korea	253 store managers (99% men) rated employees' sales-directed behaviours; 1,611 (54% men) non-managerial sales employ-ees rated the EI of their own store managers	Managerial EI was related to sales-directed behaviour of the staff, which in turn was related to objective store performance	EI was significantly related to sales-directed behaviour ($\beta = 0.15$) and while holding EI constant, sales-directed behaviour signifi-cantly predicted objective store performance ($\beta = 0.26$)	Store manager's EI was related to store cohesiveness, which in turn was related to the sales-directed behaviour of the front-line employees, which ulti-mately predicted the objective performance of the stores	Store location is not controlled; limited generalis-ability; cross-sectional design	D

upon subjective ratings by managers, and occasionally, ratings by peers/ co-workers. A few EI studies take a more expansive view of job performance and include measures of OCB. The best available evidence (from two meta-analyses and by examining the pattern of results in Table 1.1) for the relationship between EI and job performance is that the effect size is about $r = 0.28$. This is considered a "moderate" relationship by commonly accepted benchmarks of effect sizes (Cohen, 1977).

Some cautions must be observed in placing confidence in the observed relationships between EI and job performance found in the literature. For example, we concluded the following from our assessment of the studies of the relationship between EI and job performance: (1) Most studies use relatively small sample sizes, most use convenience samples, and typically sample only one group or type of employees. (2) Most studies have limited generalisability; however, (3) even though small convenience samples are typical, when you take into account all of the studies, we found diversity in job types and people across a broad variety of countries. So, while individual studies may have limited generalisability, the collection of studies as a whole provides broader external validity. (4) Typically, EI studies use cross-sectional research designs with one-time surveys. A few studies measured independent and dependent variables at two different times. Consequently, inferences about causality must be viewed cautiously. (5) Most EI studies modelled the relationship between EI and job performance as linear, but one study (Bozionelos & Singh, 2017) found a U-shaped relationship. (6) Many studies are vulnerable to common method bias. (7) Several studies found effects for the relationship between supervisor's EI on followers' job performance. (8) While many studies assess the effects of overall EI on job performance, a number of studies also found significant effects for individual dimensions of EI on job performance. (9) A small number of studies used a predictive validation design providing stronger evidence for causality and eliminating alternative explanations.

Does EI incrementally predict job performance when measures of personality and cognitive intelligence are included as predictors?

There is consistent evidence that EI demonstrates incremental validity over other predictors of job performance. Kluemper, DeGroot, and Choi (2013) found that EI consistently demonstrates incremental validity and is the strongest predictor of task performance and individually directed OCB. Ono, Sachau, Deal, Englert, and Taylor (2011) found that EI predicts some aspects of criminal investigator performance better than the five personality dimensions and cognitive ability. Law, Wong, Huang, and Li (2008) found that EI is a significant predictor of job performance beyond the effect of the General Mental Ability (GMA) battery on performance. Sy, Tram, and O'Hara (2006) found employees' EI positively predicts job performance after controlling for the Big Five personality factors. Blickle et al. (2009) found that EI explained additional variance in overall job performance ratings beyond GMA and personality traits.

What mediates the relationship between EI and job performance?

EI seems to affect job performance through several possible mediators (see Table 1.2).

For example, one study found that the relationship between EI and job performance was mediated by teamwork effectiveness in a team environment (Farh, Seo, and Tesluk, 2012). Another study found that integrating conflict management style mediated the relationship between EI and job performance (Shih & Susanto, 2010). Two studies (Carmeli, 2003; Weinzimmer, Baumann, Gullifor, & Koubova, 2017) found that the relationship between EI and job performance was mediated by its effects on work–family conflict and work–family balance.

Several studies examined mediators of particular dimensions of EI. For example, one study found that the dimension regulation of emotion was mediated by job satisfaction in influencing contextual performance (Vidyarthi, Anand, & Liden, 2014). That study also found that job satisfaction mediated the relationship between the EI dimension use of emotion and contextual performance. Another study found that the relationship between the EI dimension use of emotion and work performance was mediated by its influence in lowering burnout (Huang, Chan, Lam, & Nan, 2010). One study found that store managers' EI was related to store cohesiveness, which in turn was related to the sales-directed behaviour of the frontline employees, which ultimately predicted the objective performance of the stores (Wilderom, Hur, Wiersma, van den Berg, & Lee, 2015).

Table 1.2 Mediators and moderators of the EI→job performance relationship

Mediators	Moderators
Store cohesiveness; self-directed behaviour	Number of persons appraised
Work–family conflict	Task interdependence
Work–family balance	Power distance
Job satisfaction	Managerial work demands
Teamwork effectiveness	Cognitive intelligence
Burnout	
Integrating conflict management style	

What moderates the relationship between EI and job performance?

One study found that EI dimensions interacted with regard to job performance, such that appraising the emotions of one person was more effective than appraising the emotions of two persons (other and self), and appraising others' emotions was more effective when one's own emotions were also used or regulated. Kim and Liu (2017) found that emotional competence moderates the relationship between taking charge and employees' job performance, such that the relationship is positive only when emotional competence is high. Vidyarthi, Anand, and Liden (2014) found that relationships between leaders' emotion perceptions and employees' job performance was strengthened by task interdependence and attenuated by power distance. Farh, Seo, and Tesluk (2012) found that job context moderates the positive relationship between EI and teamwork effectiveness, such that the relationship is stronger for employees working in job contexts with higher rather than lower managerial work demands (MWD). Huang, Chan, Lam, and Nan (2010) found that one of the EI dimensions moderated the relationship between LMX and work performance such that the link between LMX and work performance was stronger for service workers with higher levels of use of emotion.

Discussion and implications for scientists and engineers

It seems logical to infer that in almost all work settings, individuals have to cooperate with others and do at least some group work tasks (O'Boyle, Humphrey, Pollack, Hawver, & Story, 2011). Furthermore, jobs in the service sector—especially those involving interactions with customers—seem likely candidates for hiring employees with high levels of EI. And, while some scientists and engineers may function as "lone rangers," fulfilling their job requirements with little or no interpersonal interaction, it is hard to imagine many of those types of jobs in today's team-oriented work environment. Consequently, working with others in science and engineering project teams requires not only heavy doses of science, technology, engineering, and mathematics (STEM) knowledge and skills; it also requires the ability to work and cooperate with others, share knowledge, and combine complementary and supplementary skill sets in the production of outcomes. Farh, Seo, and Tesluk (2012) explained that someone with high emotional perception is more likely to recognise when a team member is under significant stress by observing the team member's emotional cues and, as a result, to assist that team member by offering to off-load some responsibilities and/or make extra efforts to coordinate work activities.

One intriguing study has implications for what organisations might expect from using EI in selection and training for scientists and engineers. Cote and Miners (2006) proposed and tested a compensatory model of EI. Compensatory models propose that a specific ability predicts performance more strongly in a person who lacks other abilities than in a person who has other abilities that are related to performance. Cote and Miners (2006) assert that if compensatory effects exist, EI should predict job performance only some of the time, depending upon other abilities that individuals possess. More specifically, they proposed that cognitive intelligence moderates the association between EI and job performance, such that the association becomes more positive as cognitive intelligence *decreases*. Thus, if true for scientists and engineers, EI would be more beneficial to those who are lower in cognitive intelligence. This research needs replication in order to provide stronger evidence for organisations to act upon however.

Another study also offers insights into when EI might be important for scientists and engineers. Using trait activation theory (TAT), Farh, Seo, and Tesluk (2012) proposed that traits (in this case EI) more strongly predict trait-relevant behaviour when organisational contexts contain trait-relevant cues, which in turn activate individuals' traits and cause individuals to behave in ways that are consistent with their standing on those traits. These researchers proposed that the relationship between EI and job performance is strengthened (moderated) in job contexts involving high MWD, or jobs requiring the management of diverse individuals, functions, and lines of business. Therefore, scientists and engineers who are promoted to managerial positions will likely benefit from having higher EI.

Conclusion

In conclusion, we have conducted a systematic, evidence-based review of the relationship between EI and job performance. We have found that there is a moderate size effect for that relationship, approximately $r = 0.28$. Furthermore, EI provides incremental validity in predicting job performance beyond cognitive intelligence and personality factors. While there is a significant relationship between EI and job performance, that relationship can be moderated by a number of factors. Furthermore, the EI and job performance relationship is mediated by a number of factors as well. This information is useful to organisational decision-makers who contemplate measuring EI and using it for decision-making, such as hiring, training, and other human resource activities.

With respect to scientists and engineers, EI seems likely to be an important factor especially in contexts where there is a team environment, such as research and development teams. Furthermore, scientists and engineers who are put into management and leadership positions will likely be more effective if they have higher EI. The only scientists and engineers who may not benefit as much from higher levels of EI are those who work solo and have high cognitive intelligence. But we suspect there are very few of those employees in today's workplace.

Overall, EI seems to be an important trait with a growing evidence base for its utility in organisations. However, organisational decision-makers will be wise to acquaint themselves with that evidence base and factor that knowledge into how they apply it.

References

Barends, E., Rousseau, D. M., & Briner, R. B. (Eds.). (2017). *CEBMa guideline for rapid evidence assessments in management and organisations*, Version 1.0. Center for Evidence Based Management, Amsterdam. Available from www.cebma. org/guidelines/

Blickle, G., Momm, T. S., Kramer, J., Mierke, J., Liu, Y., & Ferris, G. R. (2009). Construct and criterion-related validation of a measure of emotional reasoning skills: A two-study investigation. *International Journal of Selection and Assessment, 17*(1), 101–118.

Bozionelos, N., & Singh, S. K. (2017). The relationship of emotional intelligence with task and contextual performance: More than it meets the linear eye. *Personality and Individual Differences, 116*, 206–211.

Carmeli, A. (2003). The relationship between emotional intelligence and work attitudes, behavior and outcomes: An examination among senior managers. *Journal of Managerial Psychology, 18*(8), 788–813.

Christiansen, N. D., Janovics, J. E., & Siers, B. P. (2010). Emotional intelligence in selection contexts: Measurement method, criterion-related validity, and vulnerability to response distortion. *International Journal of Selection and Assessment, 18*(1), 87–101.

Clarke, N., & Mahadi, N. (2017). Differences between follower and dyadic measures of LMX as mediators of emotional intelligence and employee performance, well-being, and turnover intention. *European Journal of Work and Organisational Psychology, 26*(3), 373–384.

Cohen, J. (1977). *Statistical power analysis for the behavioral sciences*. New York, NY: Routledge.

Cote, S., & Miners, C. T. (2006). Emotional intelligence, cognitive intelligence, and job performance. *Administrative Science Quarterly, 51*(1), 1–28.

Dulewicz, V., Higgs, M., & Slaski, M. (2003). Measuring emotional intelligence: Content, construct and criterion-related validity. *Journal of Managerial Psychology, 18*(5), 405–420.

Farh, C. I., Seo, M. G., & Tesluk, P. E. (2012). Emotional intelligence, teamwork effectiveness, and job performance: The moderating role of job context. *Journal of Applied Psychology, 97*(4), 890.

Greenidge, D., Devonish, D., & Alleyne, P. (2014). The relationship between ability-based emotional intelligence and contextual performance and counterproductive work behaviors: A test of the mediating effects of job satisfaction. *Human Performance, 27*(3), 225–242.

Harzing, A. W. (2018). Journal quality list. Retrieved from https://harzing.com/download/jql-2018-07-subject.pdf

Huang, X., Chan, S. C., Lam, W., & Nan, X. (2010). The joint effect of leader–member exchange and emotional intelligence on burnout and work performance in call centers in China. *The International Journal of Human Resource Management, 21*(7), 1124–1144.

Joseph, D. L., Jin, J., Newman, D. A., & O'Boyle, E. H. (2015). Why does self-reported emotional intelligence predict job performance? A meta-analytic investigation of mixed EI. *Journal of Applied Psychology, 100*(2), 298.

Joseph, D. L., & Newman, D. A. (2010). Emotional intelligence: An integrative meta-analysis and cascading model. *Journal of Applied Psychology, 95*(1), 54.

Kim, T. Y., & Liu, Z. (2017). Taking charge and employee outcomes: The moderating effect of emotional competence. *The International Journal of Human Resource Management, 28*(5), 775–793.

Kluemper, D. H., DeGroot, T., & Choi, S. (2013). Emotion management ability: Predicting task performance, citizenship, and deviance. *Journal of Management, 39*(4), 878–905.

Law, K. S., Wong, C. S., Huang, G. H., & Li, X. (2008). The effects of emotional intelligence on job performance and life satisfaction for the research and development scientists in China. *Asia Pacific Journal of Management, 25*(1), 51–69.

Law, K. S., Wong, C. S., & Song, L. J. (2004). The construct and criterion validity of emotional intelligence and its potential utility for management studies. *Journal of Applied Psychology, 89*(3), 483.

Lindebaum, D. (2013). Does emotional intelligence moderate the relationship between mental health and job performance? An exploratory study. *European Management Journal, 31*(6), 538–548.

Marler, J. H., & Boudreau, J. W. (2017). An evidence-based review of HR Analytics. *The International Journal of Human Resource Management, 28*(1), 3–26.

O'Boyle, E. H., Jr., Humphrey, R. H., Pollack, J. M., Hawver, T. H., & Story, P. A. (2011). The relation between emotional intelligence and job performance: A meta-analysis. *Journal of Organisational Behavior, 32*(5), 788–818.

Ono, M., Sachau, D. A., Deal, W. P., Englert, D. R., & Taylor, M. D. (2011). Cognitive ability, emotional intelligence, and the big five personality dimensions as predictors of criminal investigator performance. *Criminal Justice and Behavior, 38*(5), 471–491.

Pekaar, K. A., van der Linden, D., Bakker, A. B., & Born, M. P. (2017). Emotional intelligence and job performance: The role of enactment and focus on others' emotions. *Human Performance, 30*(2–3), 135–153.

Salovey, P., & Mayer, J. D. (1990). Emotional intelligence. *Imagination, Cognition and Personality, 9*(3), 185–211.

Sastre Castillo, M. Á., & Danvila Del Valle, I. (2017). Is emotional intelligence the panacea for a better job performance? A study on low-skilled back office jobs. *Employee Relations, 39*(5), 683–698.

Semadar, A., Robins, G., & Ferris, G. R. (2006). Comparing the validity of multiple social effectiveness constructs in the prediction of managerial job performance. *Journal of Organisational Behavior, 27*(4), 443–461.

Shih, H. A., & Susanto, E. (2010). Conflict management styles, emotional intelligence, and job performance in public organisations. *International Journal of Conflict Management, 21*(2), 147–168.

Sy, T., Tram, S., & O'Hara, L. A. (2006). Relation of employee and manager emotional intelligence to job satisfaction and performance. *Journal of Vocational Behavior, 68*(3), 461–473.

Vidyarthi, P. R., Anand, S., & Liden, R. C. (2014). Do emotionally perceptive leaders motivate higher employee performance? The moderating role of task interdependence and power distance. *The Leadership Quarterly, 25*(2), 232–244.

Weinzimmer, L. G., Baumann, H. M., Gullifor, D. P., & Koubova, V. (2017). Emotional intelligence and job performance: The mediating role of work-family balance. *The Journal of Social Psychology, 157*(3), 322–337.

Wilderom, C. P., Hur, Y., Wiersma, U. J., van den Berg, P. T., & Lee, J. (2015). From manager's emotional intelligence to objective store performance: Through store cohesiveness and sales-directed employee behavior. *Journal of Organisational Behavior, 36*(6), 825–844.

chapter two

Emotional Intelligence
Telling the history and discovering the models

Ana Lúcia Rodrigues and Carolina Machado
University of Minho

Contents

Introduction

For decades, it was believed that there was a positive relationship between people's Intelligence Quotient (IQ) and their performance, so that smart people were perceived to be better when compared to less intelligent people. However, IQ ignores areas such as physical aptitude, knowledge, and other skills that may result in significant advantage (Gondal & Husain, 2013). There is an important difference between IQ and EQ (Emotional Quotient). EQ can be defined as the ability of individuals to interact with others and deal with social situations (Priya & Panchanatham, 2014). Today, it seems that people who have a good balance between IQ and EQ are often more successful.

The EQ does not refer to tactics or diplomacy, but rather to the way we learn, recognise, and express our feelings and how we respond (effectively) to others when it becomes necessary to deal with this type of information (Priya & Panchanatham, 2014). It corresponds to a set of competencies that facilitate academic and professional performance, good social interactions, and a healthy life, leading to personal satisfaction (Priya & Panchanatham, 2014). This means that Emotional Intelligence (EI) is a very important concept for organisational effectiveness.

In the past, emotion was considered to interfere in rational and logical thinking. The immaturity and confusion related to emotion should be excluded if the most important thing is to think clearly. Of course, unmanaged emotions can be devastating, but as we learn more about how the cognitive and emotional sides work together, it seems to become more evident the important role that EI plays in the workplace and, in general, in organisations (Cherniss & Adler, 2000).

In recent years, emotion has begun to be seen as adaptive, useful, and functional. It organises thinking, facilitates identification of the focus of attention, and motivates behaviour (Randall, 2014). The larger and more significant role of EI in job performance, leadership, and other parts of organisational life has increased the validity of this concept (Gondal & Husain, 2013).

As organisations are spaces where problems arise and people work in a circuit of interactions with others, emotional information plays a vital role in people's lives (both professionally and personally). Sometimes, smart people who have a brilliant academic background are not good at social interaction and interpersonal relationships. EI can really be a critical factor in the effective performance of employees in the workplace.

As such, the emotionally intelligent manager can succeed under pressure, analyse problems and solve them creatively, make decisions, and manage a diverse workforce, helping the team to clarify issues and solve conflicts. The emotionally intelligent salesperson can overcome barriers to achieve goals and to readdress efforts towards more positive results. The emotionally intelligent worker is a team member who likes to go to work, proud of himself on his dedication to excellent job performance (Gondal & Husain, 2013). EI can affect all the actors in the organisation, raising the level of performance.

Organisations need leaders and employees who support the economic instability and take advantage of the opportunities that come with it. This means that human resource management must constantly rethink the systematic employment process or the approach that can provide a workforce able to take advantage of changes. Emotionally intelligent individuals are more aware of their strengths and weaknesses, which yields benefits for organisations that want to gain competitive advantage through flexibility and innovation (Goleman, 2012). Emotionally intelligent people find it easier to recognise, process, and deal with their and others' emotions in an effective and efficient way. EI provides valuable benefits to organisations that are constantly changing, improving efficiency in workers, in teams, and in organisations as a whole (Ljungholm, 2014).

Many researchers state that organisations should focus on employees' EI in order to be successful (e.g., Gondal & Husain, 2013; Priya & Panchanatham, 2014).

Emotional Intelligence

In recent years, one of the most provoking ideas in the literature on Management focuses on the possibility that a form of intelligence in the area of emotions relates to the performance of organisational members (Côté & Miners, 2006). The theory in the area of EI postulates that emotions have an informational value that makes thinking more intelligible (Brackett, Rivers, Shiffman, Lerner, & Salovey, 2006).

The concept of EI has received several definitions over the last few decades. There are several views and definitions of EI, given its breadth and complexity. Much has been said about the nature of EI, but many arguments have not yet been duly substantiated. It is clear that the research is based on an evolving EI construct that requires more rigorous studies. Nevertheless, EI's origins lie in a century of research in the fields of Psychology, Sociology, Human Development, and other disciplines.

Emotional Intelligence: history and definitions

EI has been considered one of the most controversial constructs in the behavioural and social sciences. Although its popularity has only emerged in recent years, literature recounts, in a non-consensual way, the history of this concept rewinding until several centuries before Christ.

The idea that emotions and reason are interconnected has its origin in the writings of Aristotle. The Greek philosopher assumed that passions were motivating agents of human behaviour, including *approach* and *avoidance* behaviours (Roberts, MacCann, Matthews, & Zeidner, 2010). Also the philosopher Spinoza (1677, cited by Sharma, 2008), in the seventeenth century, argued that the measurement of cognition was grounded in emotion and intellect. Cognition would be organised into three fundamental layers: emotional cognition, intellectual cognition, and a kind of intuition. For these scholars, to study emotions would be central if the goal were to understand human behaviour.

Early publications on EI began to emerge in the twentieth century with Edward Thorndike's work on Social Intelligence in 1920 (Bar-On, 2006). Thorndike subdivided Social Intelligence into EI and Motivational Intelligence (Sharma, 2008). Originally he distinguished Social Intelligence from other types of intelligence, defining it as *"the capacity to understand men and women, boys and girls, acting wisely in human relations"* (Salovey & Mayer, 1990, p. 187).

According to Bar-On (2006), many of the early studies sought to describe, define, and evaluate socially competent behaviour. Edgar Doll in 1935 unveiled the first instrument of measurement of socially intelligent behaviour for children (Bar-On, 2006). Possibly influenced by Thorndike and Doll, David Wechsler (Bar-On, 2006) included two sub-scales ("Understanding" and "Picture Arrangement"), in his well-known cognitive intelligence test that appeared to have been designed to measure aspects of Social Intelligence. One year after the first publication of this test, in 1939, Wechsler describes the influence of *"non-intellective factors"* (Bar-On, 2006, p. 13; Sharma, 2008) on intelligent behaviour, which emerges today as a reference to the concept of EI (Bar-On, 2006). In the first of many publications, Wechsler asserts that intelligence models would not be complete as long as these factors were not adequately described. He viewed EI as an integral part of the individual's personality development (Sharma, 2008). According to Sharma (2008), in 1948, Leeper introduced the idea that *"emotional thinking"* contributed to *"logical thinking."* Few years later, Mowrer (1960, cited by Sharma, 2008) pointed emotions as a higher-order intelligence.

The term Emotional Intelligence itself has been unintentionally used in the 1960s in the context of literary criticism (Van Ghent, 1961 cited by Mayer, Salovey, & Caruso, 2004) and in psychiatry (Leuner, 1966 cited by Mayer et al., 2004). According to Mayer and colleagues (2004), about 20 years later, in 1986, the expression Emotional Intelligence was used more extensively in a dissertation of Payne.

Gradually, researchers begin to shift their focus of attention, leaving aside concerns about describing and assessing Social Intelligence, turning to the understanding of the interpersonal behaviour and the role it would play in the effective adaptability of the individual (Zirkel, 2000 cited by Bar-On, 2006). This line of research contributed to the definition of human effectiveness from the social point of view, as well as strengthened an important aspect of Wechsler's definition of intelligence: *"The capacity of the individual to act purposefully"* (1958, cited by Bar-On, 2006, p. 13). Moreover, it helped the *"positioning"* of Social Intelligence as part of General Intelligence.

Commonly, it is assumed that EI derives from the concept of Social Intelligence that was first identified by Thorndike in 1920 (Roberts et al., 2010; Whiteoak & Manning, 2012; Yan-Hong, Run-tian, & Wang, 2009; Zeidner, Matthews, & Roberts, 2004). For Roberts et al. (2010), Thorndike, in his definition of Social Intelligence, advocated the idea of understanding and managing motives and emotions, and acting intelligently in human relationships. However, the difficulty in empirically distinguishing Social Intelligence from cognitive intelligence indicated that social–emotional capacity was not contemplated in other intelligence measures, with two exceptions: those proposed by Guilford and Gardner.

First, Guilford (1967, cited by Roberts et al., 2010) in his intelligence model suggests a behavioural category of intelligence, which matches the idea of dealing with emotional information.

Second, Gardner's Multiple Intelligences Theory (1994) proposes that human intelligence has multiple dimensions that must be recognised and developed through education. The author describes seven intelligences that are innate to the human being and that can be developed and improved throughout the life of the individual. They are logical–mathematical, linguistic, spatial, musical, bodily–kinesthetic, intrapersonal, and interpersonal intelligence. Given that only the last two are relevant to this chapter, we will leave aside the first five.

According to Gardner's (1994) proposal, interpersonal intelligence is centred on the individual's ability to recognise the intentions, feelings, motivations, and desires of others. It is related to the ability to deal with others, understand them and know what motivates them (Yan-Hong et al., 2009). Expressing interpersonal intelligence enables effective work with others and their understanding. On the other hand, intrapersonal intelligence entails the individual's ability to understand himself and to appreciate his own feelings, fears, motivations, and desires. In Gardner's (1994) view, this intelligence empowers the individual to better understand himself and to use this information on the regulation of his life. According to the author, an important dimension of intrapersonal intelligence includes the knowledge about other intelligences. The intrapersonal and interpersonal intelligences in Gardner's model resemble the self-knowledge, empathy, and relationships management of EI.

Also in the eighties of the twentieth century, Bar-On contributed to the concept of Emotional Quotient, from which he developed the EQ-i (Emotional Quotient Inventory), an instrument that aims to examine the conceptual model of emotional functioning (Bar-On, 2006).

Until the 1990s, the research in the area of Social Intelligence alluded to the importance of emotions for intellectual functioning, but EI was not yet understood as a concept with a psychological nature.

Initial definitions of Social Intelligence influenced the way EI was later described. The first definition of EI, made by the theorists Mayer and Salovey (1993), refers to the ability to monitor one's own emotions and to discriminate and use that information in thought and action. This proposal is followed by several others, such as Goleman's in 1995, which defines EI as the capacity to recognise emotions/feelings, to postpone satisfaction and suppress impulsiveness, to put oneself in the other's shoes and deal with other's emotions.

Three years later, in 1998, Goleman proposed the concept of Emotional Competence that he defined as a *"capacity based on emotional intelligence that results in exceptional performance at work"* (Boyatzis, 2009, p. 757).

The scientific attention to the area of EI aims to complement the traditional view of intelligence, emphasising the emotional, social, and personal contributions to intelligent behaviour. There is still a lack of consensus in the definition of EI. In spite of the profound disagreements, all theorists agree that this is an important construct for personal and professional success.

Emotional Intelligence: seminal models

In order to scrutinise EI, different perspectives are presented. The theme of EI also reaps discussion and divergence when trying to organise the perspectives on the theme. There are so many proposals of "taxonomies" for organising the approaches that are diverse in terms of the definitions and models presented. There are numerous proposals for the organisation of the perspectives on EI.

Mayer, Roberts, and Barsade (2008) distinguish between *"mental ability models"* and *"mixed models."* For these researchers, mental ability models focus on the individual's ability to process affective information, while mixed models include not only the ability to perceive, assimilate, understand, and manage emotions, but also motivational factors and affective dispositions (evident on Goleman and Bar-On's proposals). In a 2008 article, Mayer, Roberts, and Barsade position the theoretical model of Mayer and Salovey as an ability-based approach and simultaneously as an integrative approach. The authors of the article "Human Abilities: Emotional Intelligence" argue that the approach focused on specific skills is concerned with mental abilities that are important for EI. On the other hand, the integrative approach sees EI as a global aptitude. Mayer et al. (2008) mention Bar-On (2004), which points to a mixed approach, since it uses very broad definitions of EI that include *"non-cognitive capability, competence, or skill"* (p. 514), *"emotionally and socially intelligent behavior"* (p. 514), and *"dispositions from the personality domain"* (p. 514).

Bar-On (2006) understands that there are three main conceptual models: (1) the Salovey-Mayer model that defines EI as the ability to perceive, understand, manage, and use emotions to facilitate thinking; (2) Goleman's model that views EI as a broad set of skills and competencies that drive performance; and (3) the Bar-On model that describes a cross section of interrelated social and emotional skills, competencies, and facilitators that impact on intelligent behaviour.

The proposal of Mayer and colleagues and the mixed proposal of Bar-On and Goleman are consensually "categorised" in the literature. Cunha and colleagues (2007) agree with the categorisation: the first, which they referred to as Ability Models (shown in Mayer and Salovey's proposal) and the second, Mixed Models (of which the models presented by Bar-On and Goleman are illustrative). Cunha and colleagues (2007) explain that the ability models have been increasingly used and recognised in the academic and scientific context and that mixed models, which integrate mental competences and other aspects of personality or character, are the target of greater popularity.

Ability Models: Mayer, Salovey and Caruso's proposal

Salovey and Mayer (1990) view emotions as organised responses, which cross various psychological subsystems (physiological, cognitive, motivational, and experiential). For the authors, emotions are adaptive and can enrich personal and social interaction, arising as responses to internal or external events that have a positive or negative meaning for each individual. Following the authors' proposal, this positive or negative emotional reaction influences the mood/disposition and the way of acting or thinking.

According to these researchers (Mayer & Salovey, 1993), EI is a type of Social Intelligence that involves the ability to monitor one's own emotions and those of others, to distinguish them, and to use that information to guide thinking and action.

In the 1990s, Mayer, Salovey and Caruso wrote two articles about EI where they specifically defined the concept, developed a theory, and proposed an EI's measure (Mayer et al., 2004). In the view of the authors, EI operates on emotional information, which provides data about reactions and assessments that the individual uses in relationships (Mayer et al., 2004).

In the authors' initial view, EI includes the ability to monitor one's own and others' emotions, to differentiate among them, and to use this information to guide thoughts and actions. The first definition presented, *"the ability to monitor one's own and others' feelings and emotions, to discriminate among them and to use this information to guide one's thinking and actions"* (Salovey & Mayer, 1990, p. 189), was in some aspects too broad according to the authors themselves, which would give rise to interpretations according to which expressions of EI were mistaken with the capacity itself.

Thus, in 1990, the authors proposed that EI consists in a set of related mental abilities, namely: (1) evaluating and expressing emotions, in oneself and in others; (2) regulating emotions, in oneself and in others; and (3) using the emotions adaptively.

Evaluate and express emotions

According to the perspective, the processes underlying EI begin when affective information enters the perceptual system. One of the ways of emotional evaluation and expression is precisely verbal. For Salovey and Mayer (1990), *"learning about emotions depends in part upon speaking clearly about them"* (p. 191). On the other hand, affective information may lie on a non-verbal level. Much of the emotional communication is passed through non-verbal channels, which naturally hamper the task of measuring this ability.

The authors argue that emotional evaluation and expression are an important part of EI, since those who are better able to perceive and answer to their own emotions and those who express these emotions to others are emotionally smarter. These are emotional competencies since they involve the processing of emotional information within the organism, while they are fundamental to proper social functioning.

On the other hand, it is also primary the need to perceive emotions in others (and not only in oneself). Such perceptual skills ensure a more serene and interpersonal cooperation in the coherent and functional social framework of the individual. The ability to understand the feelings of others and to re-experience them in themselves appears to be related to empathy, which may be a central feature of emotionally intelligent behaviour (Salovey & Mayer, 1990). Other perspectives discussed by the authors suggest that the evaluation of others' emotions is highly related to the evaluation of one's own emotions. In fact, one does not seem to exist without the other (Salovey & Mayer, 1990). It is reasonable that the possible "illiteracy" in the analysis and treatment of one's own emotions is reflected in the analysis and treatment of the others' emotions. It will necessarily be more difficult to assess and adequately express emotions that are not treated and considered by the individual.

These competencies enable individuals to faithfully measure the affective responses of others and return the most adaptive social responses and behaviours. Those who dominate these competencies must be seen as authentic and lovable by others, as opposed to those who weakly reveal these abilities, seen by others as inattentive and rude (Salovey & Mayer, 1990).

Regulating emotions

People experience humour in a direct and introspective way. In a reflective experience, individuals have access to knowledge about one's own mood and humour. Somehow, this experience represents the willingness and ability to monitor, evaluate, and regulate emotions and mood. In this sense, it is important to analyse the mental processes underlying the ability to regulate one's own emotions (Salovey & Mayer, 1990).

Although many aspects of emotional regulation occur automatically (do not require contemplation to grieve in the face of a tragedy), some mood meta-experiences are conscious and open to questioning (Salovey & Mayer, 1990).

For the experience of emotional regulation, it is important to recognise the simultaneous occurrence of humour states and meta-experiences of humour, in multiple situations that provide individuals with information about situations that favour certain mood states. For example, if the individual experiences a pleasant state of humour while dancing, then the cause of this state of humour may be "requested" in the future, with the intention of achieving that state of humour again (Salovey & Mayer, 1990). Additionally, the individual can regulate his mood by approaching or moving away from others that cause him positive or negative moods, respectively. If you connect with others whose successes are not threatening to you, then you may experience positive affects such as pride. By contrast, if the successes of others are achieved in areas that are considered important to the individual, this may trigger negative emotional responses such as envy (Tesser, Millar, & Moore, 1988).

Salovey and Mayer (1990) stated that mood might also be directly modified, arguing that the individual has a role in the management of their mood states. It is believed that the individual is usually motivated to extend positive moods and attenuate or minimise the negative ones. The regulation of emotions can lead to more adaptive mood states.

EI also includes the ability to regulate and modify the others affective reactions. For example, an emotionally intelligent motivational speaker can rouse strong reactions in the audience, just as a jobseeker understands the favourable contribution that readiness and adjusted attire can create (Salovey & Mayer, 1990). Emotionally intelligent individuals should be experts in this process and should do so in order to achieve certain goals. On the positive side, individuals can enhance their own mood states and also the mood states of others; they can manage emotions in order to motivate others towards a rewarding end. On the negative side, manipulative and antisocial scenarios can be created, leading others to perverse and sociopathic results.

Use emotions

Individuals differ in their ability to mobilise their emotions to solve problems. Moods and emotions subtly and permanently influence some of the strategies involved in solving problems.

Flexible planning (emotions being a generating agent for the future), creative thinking (include emotional reaction to improve decision-making processes), mood redirected attention (prioritising internal and external demands and guiding the allocation of attention), and motivating emotions (since humour can be used to stimulate persistence in challenging activities) are examples of intellectual tasks where the ability to use the emotions can represent an advantage (Salovey & Mayer, 1990).

According to the model suggested by Salovey and Mayer (1990), people who have developed competences related to EI understand and express their emotions, recognise others' emotions, regulate affections, and use emotions and humour to motivate adaptive behaviour. For these researchers (Mayer & Salovey, 1993) EI remains passive while performing activities such as reading a letter's address, and is activated when the information processed involves issues of personal or emotional importance. Being able to read the non-verbal context, perceive and express emotions, and understand and regulate these emotions in ourselves is fundamental to the concept of EI (Mayer & Salovey, 1993). For Mayer and Salovey (1995), EI marks the intersection between two fundamental components of the personality: the cognitive system and the emotional system, making it possible to draw a clear distinction between EI and General Intelligence. In the article "The Intelligence of Emotional Intelligence" (1993), Mayer and Salovey argue that the underlying mechanisms of EI are different from the underlying mechanisms of General Intelligence, proposing emotionality, emotional management, and neurological connections as the three major differentiating factors. According to the authors, the general intelligence has included the concepts EI, emotionality is related to mood variations that can facilitate the prioritisation of tasks, directing attention to stimuli that need to be processed and enhancing the identification of a greater variety of responses/solutions/results. Emotional management relates to the set of mental operations that limit the emotional experience, increasing or decreasing it, and that can determine a defensive and restrictive posture or, on the contrary, an open and empathetic posture in situations of greater emotional stress. And, finally, the integration between affection and thought, which can occur in the neurological plane, concretely at the level of communication between the right and left hemispheres of the cerebral cortex. Given these differences, EI seems to differentiate general intelligence.

Considering that the first conceptual proposal was too broad and unspecific, comprising planning and creative thinking as two competences involved in the "use of emotions," the authors reformulated their theoretical model in 1997, giving greater emphasis to the cognitive components of EI, conceptualising it as a factor that enhances intellectual and emotional growth.

In the new proposal of Mayer and Salovey (1997), EI contains the ability to accurately perceive, evaluate, and express emotions; the ability to generate feelings when they facilitate thinking; the ability to understand emotion and emotional knowledge; and the ability to regulate emotions to promote growth. Based on this idea, they suggested that EI can be divided into four branches, from which results the designation of Four-Branch Model, which includes the following:

i. perceive emotions (branch 1), which explains the ability to identify feelings through stimuli (voices, stories, art, music, facial expressions,…) (Brackett et al., 2006); is considered a relatively simple psychological process and contemplates aspects such as the awareness of one's emotions and those of others, empathy and the ability to express emotions. "*Emotional perception involves recording, assessing, and deciphering emotional messages present in facial expressions, tone of voice, or cultural artifacts*" (Salovey et al., 2004, p. 449);

ii. use emotions to facilitate thinking (branch 2), that is, the ability to use the emotions in thinking and reasoning; emotions improve cognitive processing and facilitate the resolution of problems. Emotions are understood as conducting agents or triggers of cognitive activities such as reasoning, problem-solving, and decision-making. Mayer et al. (2004) explore in this dimension how emotion affects the cognitive system, and can be effectively exploited in complex mental exercises such as problem-solving, reasoning, decision-making, and creativity;

iii. understand emotions (branch 3), refers to the ability to understand the emotional evidences; focuses on the understanding of emotions and feelings, the relationship between them and how they evolve according to the context in which they are lived. It evaluates emotional transitions and allows introspection based on the observation of feelings in others. Mayer et al. (2004, p. 450) argue that this dimension *"concerns the ability to label emotions through words and to recognize relationships between examples of the affective lexicon."* Individuals who possess this ability are provided with a vast and rich emotional vocabulary, which enables them to improve interpersonal relationships while, at the same time, favouring personal and professional performance.

iv. manage emotions (branch 4), that is, the ability to deal with one's feelings and those of others, plays a central role not only in regulating emotions for their own benefit but also in establishing satisfactory relationships with others (Mayer et al., 2000; Roberts et al., 2010). The presence of this competence allows the individual to resort to strategies of maintenance and repair of their well-being, moving away harmful feelings and causing unpleasant situations while seeking to develop behaviours that lead to desired feelings. For Salovey, Mayer, and Caruso (2004, p. 450), *"the emotionally intelligent individual manages to repair his negative mood and emotions and maintain positive moods and emotions when it is appropriate to do so."*

The four levels of the model point to a hierarchy, a complexity growth of emotional abilities, according to the degree of integration of EI with other psychological subsystems: from the first level to the fourth level, from perception to management. EI fits into a process of development, starting with perception, as a basic psychological process, and ending with effective emotional regulation, as the most advanced and complex process (Koubova & Buchko, 2013; Mayer et al., 2004). Mayer et al. (2004) postulate that the valid design of EI includes the ability to process sophisticated information about one's own emotions, and the ability to use this information as a guide to thinking and behaviour. Thus, individuals with high EI pay attention, use, understand, and manage emotions, and these competencies may potentially benefit themselves and others (Mayer et al., 2004; Salovey & Grewal, 2005).

In the model presented, it is evident the integration of the emotional component with the cognitive component, with great importance being attached to the feelings and the emotions in the rationalisation of the behaviours. Several researchers argue that this model is the *"most accept-able theoretical model and therefore an appropriate model for discussion and hence for use in the mapping of biological processes"* (Tarasuik, Ciorciari, & Stough, 2009, p. 308). Pérez, Petrides, and Furnham (2005) refer to the competency model as a subjective model, since the dimensions described above depend heavily on emotional experience, and are considered unclear with regard to established criteria.

Mixed Models: Bar-On's proposal

Mixed models differ from ability models in that EI does not only relate to a set of mental competencies, but also to aspects involving personality and character (Zeidner, Matthews, & Roberts, 2004). Following this theory, will be explored two of the most discussed mixed models: Bar-On and Goleman.

Bar-On's original theoretical proposal was published in *The Bar-On Emotional Quotient Inventory (EQ-i): a test of emotional intelligence*, in 1997 (Bharwaney, Bar-On, & MacKinlay, 2007).

In 1997, Bar-On characterised EI as *"a non-cognitive matrix (…) of skills, competencies and abilities that influence a person's ability to succeed in dealing with environmental demands and pressures"* (Roberts et al., 2010, p. 823). According to Bar-On and colleagues (2000), the definition of EI proposed by Bar-On, as a non-cognitive intelligence, seems to be the most inclusive and comprehensive. Bar-On (2006) argues that it will be wiser to apply the term "Emotional-Social Intelligence," as it is composed of a set of intrap-ersonal and interpersonal skills, abilities, and facilitators that, when combined, determine effective human behaviour.

Bar-On (2006) states in the article "The Bar-On Model of Emotional-Social Intelligence (ESI)":

> emotional-social intelligence is a cross-section of interrelated emotional and social competencies, skills and facilitators that determine how effectively we understand and express ourselves, understand others and relate with them, and cope with daily demands. (p. 15)

As discussed by Bar-On, Brown, Kirkcaldy, and Thomé (2000), EI is a non-cognitive intelligence defined as an array of emotional, personal, and social skills that influence the individual's ability to effectively deal with environmental demands and pressures.

For Bar-On et al. (2000), the key factors involved in this model include: (1) intrapersonal skills: ability to know and understand oneself and one's own emotions, express one's own feelings and ideas; (2) interpersonal skills: ability to understand and appreciate the feelings of others as well as to establish and maintain mutually responsible and satisfying relationships; (3) adaptability: ability to check the feelings of others through objective external signals and accurately assess the immediate situation, flexibly change emotions and feelings in the light of the changes that occur in a given situation, and solve personal and interpersonal problems; (4) stress management strategies: ability to cope with stress and control strong emotions; and (5) motivational and general mood factors: ability to be optimistic, to appreciate oneself and others, and to feel and express positive feelings (Bar-On et al., 2000).

Each of the five key factors presented above includes a number of skills, competencies, and facilitators that are closely related. The description of the skills and competences aggregated to each of the factors is described in Table 2.1.

Table 2.1 Skills presented in the Mixed Model proposed by Bar-On

Key factor	Habilities and skills	Description
Intrapersonal	Self-esteem	Perceiving, understanding, and accepting oneself
	Emotional self-awareness	Be aware and understand emotions themselves
	Assertiveness	Express yourself effectively and constructively, yourself and your emotions
	Independence	Be self-reliant and free from emotional dependence on others
	Self-actualisation	Strive to achieve personal goals and achieve your potential
Interpersonal	Empathy	Be aware and understand how others feel
	Social responsibility	Identify yourself with a social group and cooperate with others
	Interpersonal relationships	Establish mutually satisfactory relationships and relate well with others
Stress management	Stress tolerance	Manage emotions constructively and effectively
	Impulse control	Control emotions constructively and effectively

(Continued)

Table 2.1 (Continued) Skills presented in the Mixed Model
proposed by Bar-On

Key factor	Habilities and skills	Description
Adaptability	Problem-solving	Solve problems of a personal or interpersonal nature in an effective way
	Reality testing	Objectively validate thoughts and feelings with external reality
	Flexibility	Adapt and adjust feelings and thoughts to new situations
General mood	Optimism	Be positive and look at the bright side of life
	Happiness	Feel satisfaction with yourself, with others, and with life in general

Font: Bar-On (2006).

The dimensions presented join emotion and personality. The dimension of intrapersonal intelligence encompasses components alluding to self-knowledge, as the way each person recognises himself, how he interprets and understands his emotional states, and assertiveness, which is understood as the facility to express feelings or internal states. The second dimension focuses on the skills needed for social interaction and how the individual handles the emotion of others. It unfolds into components ranging from empathy (understanding of others' feelings) to responsibility for others' feelings.

Stress management refers to how the individual looks at and overcomes any adverse situations that cause emotional stress as well as the ability to control impulses and regulate emotions for his own benefit. The fourth dimension—adaptability—refers to the skills that are called for resolving and overcoming adversities, problems, and changes. These situations require emotions that are facilitators of thinking and reasoning and that facilitate adaptation to the demands of the environment.

The fifth and last dimension, general mood, is based on happiness and optimism that are seen as reflecting the degree of satisfaction of the individual with his or her life. It is expressed in the positive emotions and/or states of well-being that promote success.

In the 1980s, Bar-On developed an instrument for measuring ESI, the EQ-i, so that it could examine the conceptual model of emotional and social functioning (Bar-On, 2006). In the 1990s, Bar-On attempted to test the factorial structure of his model through that instrument. From the analysis, some differences (theoretical structure and that obtained in the exploratory factorial analysis) emerged and that led to a review of the 15 factors initially proposed (as shown in Table 2.1: self-esteem, emotional self-awareness, assertiveness, independence, self-actualisation, empathy, social responsibility, interpersonal relationships, stress tolerance, impulse control, problem-solving, reality testing, flexibility, optimism, and happiness).

The results suggested a structure of ten factors, which Bar-On (2006, p. 8) considered *"empirically feasible and theoretically acceptable"*: self-esteem, interpersonal relationship, impulse control, problem-solving, reality testing, stress tolerance, assertiveness, and empathy. The author concludes (Bar-On, 2006, p. 8):

> These ten factors appear to be the key components of ESI, while the five factors that were excluded from the second confirmatory factor analysis (Optimism, Self-Actualization, Happiness, Independence and Social Responsibility) appear to be important correlates and facilitators of this construct.

The suppression of the factors Optimism and Happiness leads to the disappearance of the fifth key factor, the "General Mood." As explained by Bar-On (2006), the reason why these five factors were excluded from the analysis is justified by the fact that the items that constitute each of these factors saturate strongly in other factors under analysis. If, on the one hand, it is easy to understand the factorial construction underlying the theoretical model proposed by Bar-On, given that the ten factors participate in the set of emotional and social capacities that compose the ESI, it remains to explain the "permanence" of the five factors, and these are now called facilitators.

Mixed Models: Goleman's proposal

In the 1990s, the book "Emotional Intelligence" written by Goleman (1995) reinforces the public interest in EI (Roberts et al., 2010). Daniel Goleman, author of several books on EI (including "Working with Emotional Intelligence," released in 1998), considered that EI refers to the ability to know and manage emotions, to motivate oneself, to recognise emotions in others, and to deal with relationships (Goleman, 2012). Goleman (2012), in his book "Working with Emotional Intelligence," links EI with *"competences associated with self-awareness, self-control, social awareness and relationship management"* (p. 823). Goleman (2012) claims to have adapted the model of Salovey and Mayer to a version he deems most useful in understanding how the talents associated with EI influence the working life.

Goleman (2012) postulates that EI underlies emotional competence and that emotional competence is a necessary antecedent of good performance. EI improves the potential of the employee to learn, and emotional competence translates that potential into the task domain. Demonstrating high EI is insufficient. EI, as a trait, only suggests that the worker has the ability to learn new skills, not that these skills have already been learned. To broaden the reader's understanding, Goleman uses a musical analogy: although some individuals were born with singing aptitude, those who do not receive vocal training will never shine like opera singers. In this line, only emotionally intelligent workers with strong social skills (i.e., with emotional competence) will be able to build relationships and solve conflicts (Young, Arthur, & French, 2000).

Goleman (2012) believes that emotional skills are more important to personal success than people's cognitive, social, and professional skills. He argues that EI brings together independent competencies that give unique contributions to job performance, and are a necessary but not sufficient condition for effective leadership or cooperative skills.

In the book *Emotional Intelligence—The Revolutionary Theory That Redefines What It Is to Be Intelligent*, Goleman (1995) discusses the importance of emotions in human life. Based on the principle of evolution, the author suggests that the existence of emotional centres in the brain (e.g., limbic system) precedes rational centres. This evidence, according to the author, is supported by the principle of survival inherent in any living organism. In this sense, the emotional side of the brain is the one that tells the body the answer to adopt in the face of the oncoming challenge: to fight or to run away, to ignore or to accept. In Goleman's (1995) argument, *Homo sapiens'* neocortex, responsible for rational thinking and the ability to analyse and plan the most appropriate responses, is a secondary development that adds to humans what makes them distinct from other animals. According to Goleman (2012, p. 323), *"These two different types of intelligence—the intellectual and the emotional—express the activity of different regions of the brain."*

Despite weak scientific support, in 1996, Daniel Goleman stated that EI is twice as important as IQ for success in life and made it an everyday concept (Priya & Panchanatham, 2014). In 1997, Goleman presented a mixed model in which EI included skills such as self-control, zeal, persistence, and self-motivation. These skills convert into a set of emotionally intelligent skills that are defined by character. For the author, EI is the variable that most explains the success or failure of the individual, especially in work situations or in interpersonal relationships.

The model proposed by Goleman (1997) relates EI competencies to the individual's personality traits and includes intrapersonal and interpersonal skills such as emotional self-awareness (recognising one's own emotions), emotional control (managing one's feelings by adjusting oneself), using emotions for personal benefit, recognition of emotions in others (empathy of feelings), and ability in interpersonal relationships (skills in social relationship). For the author, EI *"is the ability to recognize our own*

*feelings and those of others, to motivate ourselves and to manage emotions well in
ourselves and in our relationships"* (Goleman, 2012, p. 333).

These skills play a salient role in group management (in particular in the centralisation of efforts), conflict management, and social sensitivity—the ability to perceive feelings and motives in people (Goleman, 1997).

The five dimensions of EI designated by the author (Goleman, 2012) as a set of capacities that can be strengthened throughout life are then:

i. Self-awareness—it is essential to realistically assess our own abilities, which will allow self-confidence to flourish; simultaneously, we must be able to recognise what we feel at every moment, so that we can use that information in decision-making processes;
ii. Managing Emotions—managing emotions so that the tasks to be performed are facilitated and also that emotions do not result in unwanted consequences for ourselves and others; in addition, the ability to postpone the reward for achieving the goals is important;
iii. Motivation—ability to use the deepest preferences to take the initiative and guide action towards the goals, persisting even in the face of setbacks and disappointments;
iv. Empathy—being able to perceive what other people feel, understanding and adopting their perspective, being able to cultivate affinities with a great diversity of people;
v. Social Competences—ability to analyse social situations and networks of relationships and interact harmoniously with other people, performing good emotional management, persuading and leading others, managing conflicts and differences, cooperating and working as a team.

In sum, in his first conceptual proposal, Goleman understands that EI appears "hierarchically" structured in five clusters. At the first level, the individual must understand his own state and its effect on others. This awareness will facilitate decision-making processes and enable better relations with others. At the second level comes the emotional self-regulation. This competence refers to the individual's ability to calm himself by controlling his impulses and being able to react prudently in diverse situations. The motivation refers to the individual's ability to direct emotions in order to achieve an intended goal, persisting in its accomplishment. Empathy, on the fourth level, is "oriented" to the other, is less focused on the inner world of the individual. It respects the understanding of others' emotions, their wants or needs, and the ability to understand undisclosed feelings. Finally, social skills, related to the ability to interact constructively with others, trying to manage others' emotions in order to harmonise personal relationships. Social skills close the cycle of five competencies that can be developed throughout life.

Conclusion

Currently, EI is seen as a relevant competence for the personal and professional success of individuals. There are several definitions and models of EI proposed in the literature. Globally, are seminal the Mayer and Salovey Ability Models of EI and the Mixed Models of EI of Bar-On and Goleman.

Mayer and Salovey come to EI as a group of related mental abilities and define it as the ability to monitor emotions and feelings of oneself and the emotions and feelings of others, distinguishing them and using them to guide thinking and actions. The most seminal proposal of the ability approach stresses that EI expressions should not be confused with the concept itself. Likewise, in this model, personality traits such as persistence, zeal, self-control, or other positive attributes should not be included. Emotional abilities can be seen in a growing continuum of complexity, in four branches, from perception to management. The four branches include the following abilities: (1) to perceive emotions in oneself and in others; (2) to use emotions to facilitate thinking; (3) to understand emotions, the emotional language, and the signals transmitted by the emotions; and (4) to manage emotions in order to achieve certain goals.

In this model, it is evident the integration of the emotional component with the cognitive component, with great importance being attached to feelings and emotions in the rationalisation of behaviours.

In the Bar-On's model, the Socio-Emotional Intelligence (ISE) is seen as a set of emotional and social capacities. In particular, ISE is based on five key factors: intrapersonal, interpersonal, stress management, adaptability, and general mood. It is composed of the individual's ability to understand himself and others, to express himself and to relate to others, to adapt to change and to solve social and personal problems, and above all, to deal with strong emotions. The mixed model proposed by Bar-On distances itself from the previously explored ability model insofar as it advocates ISE as a model of skills broadly associated with personality traits.

The fundamental distinction of Goleman's model from the rest is related to the role that performance plays in his model. For Goleman, EI consists of a set of five competences (self-awareness, emotional management, motivation, empathy, and social skills), which in turn function as a modelling agent and facilitator of effective performance. For Goleman, EI's relationship with the results obtained in personal and professional life is central.

References

Bar-On, R. (2004). The Bar-On Emotional Quotient Inventory (EQ-i): rationale, description and summary of psychometric properties. In *Measuring Emotional Intelligence: Common Ground and Controversy*, ed. Glenn Geher. Hauppauge, NY: Nova Science Publishers, 111–142.

Bar-On, R. (2006). The Bar-On model of emotional-social intelligence (ESI). *Psicothema*, 18 (supl.), 13–25.

Bar-On, R., Brown, J., Kirkcaldy, B., & Thomé, E. (2000). Emotional expression and implications for occupational stress: An implication of the Emotional Quotient Inventory (EQ-i). *Personality and Individual Differences*, 28 (6), 1107–1118.

Bharwaney, G., Bar-On, R., & MacKinlay, A. (2007). *EQ and the Bottom Line: Emotional Intelligence Increases Individual Occupational Performance, Leadership and Organisational Productivity*. UK: MBE Maidenhead.

Boyatzis, R. E. (2009). Competencies as a behavioural approach to emotional intelligence. *Journal of Management Development*, 28 (9), 749–770.

Brackett, M. A., Rivers, S. E., Shiffman, S., Lerner, N., & Salovey, P. (2006). Relating emotional abilities to social functioning: A comparison of self-report and performance measures of emotional intelligence. *Journal of Personality and Social Psychology*, 91 (4), 180–195.

Cherniss, C., & Adler, M. (2000). *Promoting Emotional Intelligence in Organisations: Make Training in Emotional Intelligence Effective*. Alexandria, VA: American Society for Training & Development.

Côté, S., & Miners, C. T. H. (2006). Emotional intelligence, cognitive intelligence, and job performance. *Administrative Science Quarterly*, 51 (1), 1–28.

Cunha, M. P., Rego, A., Cunha, R. C., & Cabral-Cardoso, C. (2007). *Manual de Comportamento Organizacional e Gestão*. Lisboa: Editora RH, Lda.

Gardner, H. (1994). *Estruturas da Mente - A Teoria das Inteligências Múltiplas*. Porto Alegre: Artes Médicas Sul.

Goleman, D. (1995). *Inteligência Emocional - A teoria revolucionária que redefine o que é ser inteligente*. Rio de Janeiro: Editora Objetiva, Ltda.

Goleman, D. (1997). *Inteligência Emocional*. Lisboa: Temas e Debates - Círculo de Leitores.

Goleman, D. (2012). *Trabalhar com Inteligência Emocional*. Lisboa: Temas e Debates - Círculo de Leitores.

Gondal, U., & Husain, T. (2013). A comparative study of intelligence quotient and emotional intelligence: Effect on employees' performance. *Asian Journal of Business Management*, 5 (1), 153–162.

Koubova, V., & Buchko, A. (2013). Life-work balance – Emotional Intelligence as a crucial component of achieving both personal life and work performance. *Management Research Review*, 36 (7), 700–719.

Ljungholm, D. P. (2014). Emotional intelligence in organisational behavior. *Economics, Management and Financial Markets*, 9 (3), 128–133.

Mayer, J., Caruso, D., & Salovey, P. (2000). Emotional intelligence meets traditional standards for an intelligence. *Intelligence*, 27 (4), 267–298.

Mayer, J. D., Roberts, R., & Barsade, S. (2008). Human abilities: Emotional intelligence. *Annual Review of Psychology*, 59, 507–536.

Mayer, J. D., & Salovey, P. (1993). The intelligence of emotional intelligence. *Intelligence*, 17, 433–442.

Mayer, J., & Salovey, P. (1995). Emotional intelligence and the construction and regulation of feelings. *Applied and Preventive Psychology*, 4 (3), 197–208.

Mayer, J. D., & Salovey, P. (1997). What is emotional intelligence? In *Emotional Development and Emotional Intelligence: Educational Implications*, eds. Peter Salovey, & David J. Sluyter. New York: Basic Books, 3–31.

Mayer, J., Salovey, P., & Caruso, D. (2004). Emotional intelligence: Theory, findings, and implication. *Psychological Inquiry*, 15 (3), 197–215.

Pérez, J. C., Petrides, K. V., & Furnham, A. (2005). Measuring trait emotional intelligence. In *Emotional Intelligence: An International Handbook*, eds. Ralf Schulze & Richard D. Roberts. Cambridge, MA: Hogrefe and Huber, 181–201.

Priya, A., & Panchanatham, N. (2014). Personality in relation to emotional intelligence among the professionals. *International Journal of Management and Innovation*, 6 (2), 1–14.

Randall, K. (2014). Emotional intelligence: What is it, and do Anglican clergy have it? *Mental Health, Religion & Culture*, 17 (3), 262–270.

Roberts, R., MacCann, C., Matthews, G., & Zeidner, M. (2010). Emotional intelligence: Toward a consensus of models and measures. *Social and Personality Psychology Compass*, 4 (10), 821–840.

Salovey, P., & Grewal, D. (2005). The science of emotional intelligence. *Current Directions in Psychology Science*, 14 (6), 821–840.

Salovey, P., & Mayer, J. D. (1990). Emotional intelligence. *Imagination, Cognition and Personality*, 9, 185–211.

Sharma, R. R. (2008). Emotional intelligence from 17th century to 21st century: Perspectives and directions for future research. *VISION – The Journal of Business Perspective*, 12 (4), 59–66.

Tarasuik, J. C., Ciorciari, J., & Stough, C. (2009). Understanding the neurobiology of emotional intelligence: A review. In *Assessing Emotional Intelligence: Theory, Research, and Applications*, eds. Con Stough, Donald H. Saklofske, & James D. A. Parker. Boston, MA: Springer, 307–320.

Tesser, A., Millar, M., & Moore, J. (1988). Some affective consequences of social comparison and reflection processes: The pain and pleasure of being close. *Journal of Personality and Social Psychology*, 54, 49–61.

Whiteoak, J., & Manning, R. (2012). Emotional intelligence and its implications on individual and group performance: A study investigating employee perceptions in the United Arab Emirates. *The International Journal of Human Resource Management*, 23 (8), 1660–1687.

Yao, Yan-Hong, Wang, Run-tian, & Wang, K. Y. (2009). The Influence of Emotional Intelligence on Job Performance: Moderating Effects of Leadership. *International Conference on Management Science & Engineering* (16th), 14–16 September, Moscow, Russia, 1155–1160.

Young, B. S., Arthur, W. A., Jr., & French, J. (2000). Predictors of managerial performance: More than cognitive ability. *Journal of Business and Psychology*, 15 (1), 53–72.

Zeidner, M., Matthews, G., & Roberts, R. (2004). Emotional intelligence in the workplace: A critical review. *Applied Psychology: An International Review*, 53 (3), 371–399.

chapter three

How Japanese managers use NLP in their daily work

Yasuhiro Kotera and William Van Gordon
University of Derby

Contents

Introduction

Although Neuro-Linguistic Programming (NLP) has its origin in America, it is used in numerous countries across the globe (Kotera, Sheffield, & Van Gordon, 2018). For example, NLP has been shown to improve occupational stress amongst Iranian nurses (HemmatiMaslakpak, Farhadi, & Fereidoni, 2016), reduce perceived stress and time-related pressure in Indian workers (Rao & Kulkarni, 2010), and be a popular technique for improving work effectiveness in Japan. Indeed, the NLP Connection (one of the original NLP organisations founded by Richard Bandler) has certified 1,725 practitioners, 1,321 master practitioners, 373 trainer associates, and 40 trainers since it first started teaching NLP in Japan in 2003 (Kotera & Van Gordon, 2019).

One reason underlying NLP's popularity in Japan is its focus on increasing awareness of mental health problems. Contrary to the fact that people in Okinawa have been described as one of the happiest groups of people in the world (Buettner, 2017), mental illness is a significant public health concern in Japan (Kotera, Gilbert, Asano, Ishimura, & Sheffield, 2018). Indeed, despite its fairly stable population levels, the number of depressed people in Japan has increased from 441,000 in 1999 to 1,041,000 in 2008, which corresponds to a 136% increase in less than 10 years (Ministry of Health, Labour and Welfare [MHLW], 2015). The suicide rate amongst Japanese people has been one of the highest amongst developed countries for many years such that each year, approximately 30,000 people commit suicide in Japan (Organisation for Economic Co-operation and Development, 2015). One third of all suicides in Japan are committed by workers, and in respect of suicides committed by the Japanese worker population, approximately one in three instances are due to problems at work (National Police Agency, 2016). Furthermore, 60% of Japanese people experience high levels of mental distress (MHLW, 2010), and an increasing number of workers are applying for compensation claims for mental health problems (an increase from 200 in 2000 to 1,500 applications in 2015) (MHLW, 2016).

In line with growing awareness of mental health problems in Japan (Kobori et al., 2014), the number of companies that implement mental health support programmes has increased from 23.5% in 2002 to 47.2% in 2012 (MHLW, 2013). Nevertheless, approximately half of all Japanese companies do not offer mental health support, primarily because they do not feel the necessity (MHLW, 2013). If suicides and depression in Japan were eliminated, the yearly benefit is estimated to be approximately 2.7 trillion Japanese yen, amounting to 0.7% of GDP (Kaneko & Sato, 2010).

These reports highlight the need for effective mental health interventions for Japanese workers, and they are likely to constitute one reason underlying the growing popularity of NLP amongst Japanese managers. This chapter draws on the first author's experience as an NLP researcher and practitioner and outlines real and hypothetical examples in order to explicate how Japanese managers use NLP skills in their day-to-day work. The chapter also outlines recommendations for practitioners wishing to introduce and utilise NLP approaches in their own occupational and/or healthcare settings.

Example 1: Position change

A senior manager in the HR department of a pharmaceutical company had a consultation with one of his team members (pseudonym: Ken). Ken had some conflicts with another member of staff (pseudonym: Naomi). The manager brought two chairs and utilised the 'position change' technique to help Ken understand Naomi's perspective by identifying the goals, positive intentions, interests, and concerns that Naomi might have. By undertaking this exercise, Ken gained insight into why there was conflict and how it might be solved. The feelings he had towards Naomi changed as the exercise unfolded. Indeed, before the intervention, Ken felt stressed, and did not want to see Naomi anymore, but after the intervention, he was more able to see Naomi from an attitude of accep-tance and understanding. This is a relatively easy exercise that can be implemented in the following way (Manager as the interventionist and Ken as a recipient).

1. Position two chairs opposite each other (Figure 3.1).
2. Ken sits on one of them (1st position) and imagines Naomi sitting on the other. Ken then faces Naomi's chair and the Manager iden-tifies how Ken feels about his relationship with Naomi (e.g., *Now you are facing Naomi, how do you feel, Ken?*). Ken may respond that he is stressed, angry, or upset. Noticing these negative emotions is referred to as 'tasting the poison', and it can foster a desire for change.
3. The Manager then asks Ken to express his feelings in a simple man-ner to Naomi. Ken may say *'Why do you always wait too long to send our proposals to our clients?'* or *'You always make small errors that degrade the quality of our product'* etc.
4. Once Ken's thoughts are expressed, the Manager asks Ken to stand up from the chair and imagine that he has left his body on the chair while Ken moves to a position that is in-between the two chairs (neutral position). The Manager then asks Ken *'Who sits on this chair?'* pointing to one chair at a time. Verbalising his response (e.g., *'Ken sits on that chair'* and *'Naomi sits on this chair'*) from the neutral posi-tion can help Ken to imagine the relationship more vividly, and to dissociate from the negative feelings expressed in the 1st position. The Manager should guide the exercise so that Ken's negative feel-ings expressed in the 1st position are no longer present in the neu-tral position. Useful ways to ensure that Ken has transformed any

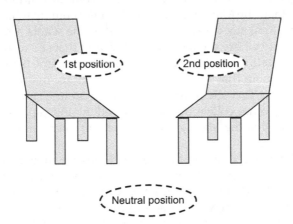

Figure 3.1 Positions used in the position change technique.

negative feelings harboured during the 1st position are (1) guiding Ken to breathe gently but deeply, (2) guiding Ken to walk around the chairs, and (3) asking Ken out-of-context questions such as what the last four digits of his phone number are, or who his favourite comedian is. In the neutral position, the Manager also can ask how the two (imaginary) individuals sitting in the chairs look as they face each other.

5. Next, the Manager asks Ken to sit on the other chair (2nd position), while guiding Ken's thoughts using a script such as *'Now I want you to sit on this chair. You see Naomi here, and as you sit on this chair, you will become Naomi. Sit like Naomi, make a posture like Naomi, make a facial expression of Naomi, and breathe like Naomi. When you get a sense of Naomi, let me know'.*

6. The Manager should wait enough time for Ken to really appreciate what it is like to stand in Naomi's shoes. When Ken is ready, the Manager can ask *'Who are you? (Naomi)'* and *'Who sits in front of you? (Ken)'*. Then, the Manager can ask *'How do you feel facing Ken now?'* Naomi (acted by Ken) may say that she is scared, anxious, or threatened.

7. As Naomi (acted by Ken) starts to express her thoughts about Ken, the Manager continues to explore what Naomi thinks about the work relationship with Ken. Naomi (i.e., acted by Ken) may say *'Why can't you trust me?'* or *'If you pressure me like that, I cannot think straight'* etc. Key questions that the Manager can ask include *'How do you want to feel about this work relationship instead?'* and *'What do you want from Ken or other colleagues?'* The Manager may also ask Naomi what her goals, positive intentions, interests, or concerns are, especially when Naomi describes specific issues. Identifying positive intentions is a key objective of NLP and thus effective communication is required. At this stage, the Manager may want to take notes of what Naomi (i.e., acted by Ken) says. Naomi may say that she wants a more relaxed relationship where she feels respected, and that she would like short regular meetings with Ken to review minor errors. Naomi (acted by Ken) may note that her goal and intention are to create a supportive team but that she is concerned about feeling unconfident given that many other colleagues seem confident and capable. Again, it is important that the Manager remains as an interventionist, focusing on listening.

8. Once the above step is complete, the Manager should ask Ken (who was acting the role of Naomi) to stand up, and return to the neutral position. The Manager can again ask who sits on each chair, making sure that the feelings expressed in the 2nd position are dissociated. Use the useful ways given in the step 4 to ensure the feelings in the 2nd position have been relinquished. The Manager once again asks

how the relationship between the two looks at this point. Ken may start to report a changed perspective.

9. Having reflected upon Naomi's possible thoughts, the Manager invites Ken to return to his original chair (i.e., the 1st position). The Manager then repeats to Ken what the (imaginary) Naomi reported. Having better understood Naomi's possible thoughts, feelings, goals, and positive intentions, the Manager then asks Ken how he feels about his relationship with Naomi now.

If necessary, the above process can be repeated to further explore the working relationship. It is also possible to invite the employee to sit in the Manager's chair in order for them to understand employee conflicts from a managerial position (in NLP, this is called the 'meta-position'). The position change process can also be used (for example) to better understand a customer or a supplier's perspective.

Example 2: Alignment of the neuro-logical levels

Another NLP skill that trained NLP Japanese senior managers sometimes employ is that of alignment with the neuro-logical levels, which were developed by Dilts (1996). This conceptual framework of six different levels was inspired by the theoretical foundations of learning and change (Bateson, 1972). The framework involves organising our experience into the following six levels: environment, behaviours, capabilities, beliefs/values, identify, and spiritual levels. Japanese managers use this framework to facilitate communication with their colleagues, and to create a team vision. The neuro-logical levels are often portrayed in a two-triangle figure; the bottom triangle represents the individual realm and the top triangle represents the social realm. Though in some cases, the lower levels could influence the higher levels, in most cases, the higher levels influence the lower levels. Therefore, it is important to determine 'for what or whom, you work (spiritual level)' and 'who you are (identity)', as the upper levels can change or alter other levels (Yamazaki, 2007) (Figure 3.2).

Identifying each level is often helpful for creating a congruent individual identify (and this can be applied to team-building too), which can lead to high performance (Dilts, 1996). For example, imagine a professional pianist or athlete that you know. Probably they are in the environment where they can access their practice facilities relatively easily (environment level), and they practise every day (behavioural level). They are able to play the piano elegantly or play a certain sport very well (capability level), and they probably have deep values-based belief relating to their profession (e.g., *'Music is so important to me'* or *'Baseball is what I live for'*—values/beliefs level). They identify themselves as a 'professional pianist' or 'professional athlete' (identity level). They may also receive

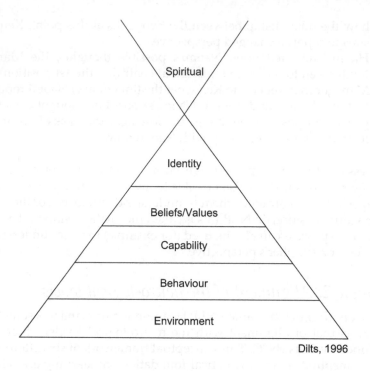

Dilts, 1996

Figure 3.2 Neuro-logical levels (Dilts, 1996).

pleasure in the social realm, and have some spiritual purposes for play-ing the piano or sport. The pianist may keep refining their skills even further in order to provide the healing/pleasurable sounds to the audi-ence, and the athlete may keep practising hard and pushing their limits in order to win the next game for their fans (spiritual level). The congru-ency they have is an essence of their high performance (Dilts, 1996). This also applies to leadership in organisations. Dilts analysed many notable leaders including Steve Jobs and Mohandas Gandhi, using the neuro-logical levels (Dilts, 1996).

A senior Japanese manager working within quality management of a leading automobile manufacturer used this framework as part of his team meetings. One aspect of his company's motto considers contributing to clients' happiness through a quality car life, which refers to the spiritual level of the neuro-logical levels. Based on this, his team discussed who they need to be or what kind of team identity they need to have, in order to provide a quality car life. Given they worked in quality management, they identified themselves as a quality management team that pursues per-fect accuracy and foresees potential problems their clients may encounter, and seeks to implement preventable solutions beforehand (identity level). The team valued attention to detail, and they believed that high quality

influences the clients' well-being (beliefs/values level). The team has frequent meetings to share best practice, including at international quality
conferences (behavioural level). Their workplace has clear signs with
visuals, and each worker's work process is visualised and transparent to
their colleagues (environment level). Through this exercise of aligning the
neuro-logical levels, the team were more able to acknowledge why they do
what they do, and this helped to increase team performance and worker
satisfaction.

The following is a method for enacting the alignment of neuro-logical
levels. This can be done alone but many people find it easier to do in the
presence of somebody who can listen to each step. The task can be done
at a desk, but it is often more powerful to do it spatially using a suitably
sized room (see Figure 3.3; four steps in front of you, and one step behind):

1. Think of your goal in the next few years. Imagine what you would
   see, hear, and feel when you achieve it. It is important here to experience the pleasure of this achievement as it can help increase focus
   upon the following steps. For example, someone may say that they
   want to be a team leader in 3 years (let's call this person 'Jane'). Jane
   may imagine herself sitting in a certain place in the office, and leading a team meeting elegantly. She might imagine hearing her team
   members or clients thanking her for her leadership. She may feel a
   sense of inner authority and fulfilment in the role of team leader.

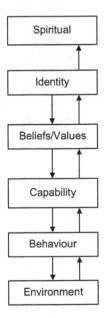

Figure 3.3 Space to be used for the alignment of the neuro-logical levels.

2. Keep feelings positive. When you achieve the goal, what is your self-image? How do you see yourself? Is there a title or nickname you would give yourself upon achieving the goal—a keyword that links to a sense of your identity? When you imagine yourself with this title or nickname, you should be able to more easily remember this positive sensation. In NLP terms, it is called an 'auditory anchor': every time you hear it (externally or internally), you should experience a certain sensation.

3. Make a step forward to the beliefs and values level. Think what you (who have achieved this goal) now value or believe. What would be important to you? What manner of thinking would you have? Ask yourself *'Why do you work?'*, because the beliefs and values level is related to our motivation, which can be clarified by asking 'why'. For example, Jane's future self as the team leader might believe that compassionate leadership is better than pressure-based leadership.

4. Now step forward to the capability level. What capabilities would you have, or be utilising predominantly when you achieve your goal? You may want to describe the proficiency of the capabilities. This process will be useful when you make plans to achieve the goal, as it can illustrate specifically what you need to improve on. For example, Jane as the team leader might be using a high level of management capabilities and assertiveness to lead her team. She may also have a high level of capability in recognising the needs of the entire organisation.

5. Next, proceed to the behaviour level. When you achieve the goal, what will be your behavioural patterns? For example, Jane might be having a regular meeting with her team members as well as with executives. She might be undertaking some management training such as assertiveness skills or team-building skills.

6. At the base of the neuro-logical levels is the environment level. Where will you be once the goal is achieved? Will you be in the same office as you are now? What about other work environment issues? How will your time management be? Will you have more autonomy over your time? In Jane's case, she might be working in the same office, but at a different desk, where she can easily engage with all of her team members. She may be working more flexibly, allowing her to maintain a better work–life balance.

7. Next, revisit each step and then enter into the spiritual level. At this level, you think for what or whom will you be working. This is important because according to NLP, our potential is more fully utilised when we apply sponsorship principles. For example, Gandhi's relentless non-violence movement was for the independence of India. Jane as the team leader might be working in order to make her

Table 3.1 Questions, targets, and roles of the neuro-logical levels

Levels	Questions	Targets	Role
Spiritual	*For what/whom*, do you work?	Vision & Purpose	Awakener
Identity	*Who* are you as a worker?	Role & Mission	Sponsor
Beliefs & Values	*Why* do you work?	Motivation & Permission	Mentor
Capabilities	*How* do you work?	Perception & Direction	Teacher
Behaviours	*What* do you do in your work?	Actions & Reactions	Coach
Environment	*When and where* do you work?	Constraints & Opportunities	Caretaker & Guide

team members and organisation happy, which would lead to client satisfaction. This would help her to create a clear vision as the team leader.

The above procedure started with the identity level, as that is what many clients find relatively easy to clarify. However, if your colleague has already arrived at clarity in respect of some of the levels, it is acceptable to start from another level. What is important is to clarify the 'experience' of the goal at each level, to foster congruent working towards the goal as well as an understanding of what is required in order to achieve it. Table 3.1 summarises how information can be elicited at each level (questions), what function each level has (targets), and what the role of each level is (role).

Example 3: Neuro-logical levels in communication

The discussion above suggests another application of the neuro-logical levels—communication (Yamazaki, 2007). Japanese managers have reported that the neuro-logical levels can help them to analyse their communication, including reviewing the effect that communication at a given level had on the employees. Being criticised at a higher level is often more psychologically painful than being criticised at a lower level. The following exercise, that requires you to read and identify with each of the below statements, explains how the neuro-logical levels apply to communication

1. 'The thing that is front of you is dangerous.'
2. 'That behaviour you take is dangerous.'
3. 'The lack of your capabilities is dangerous.'
4. 'That way of thinking you have is dangerous.'
5. 'You are dangerous.'

*Statement 3 states 'the lack of' in order to make sense with the word 'dangerous'.

**The word 'dangerous' is just an example: You can change it to 'safe' or other words.

People typically report that while they feel 'danger' externally when they hear Statements 1 and 2, they feel it more internally when they hear the latter statements. Statement 1 relates to the environmental level, Statement 2 relates to the behavioural level, Statement 3 relates to the capability level, Statement 4 relates to the beliefs/values level, and Statement 5 relates to the identity level. Statement 5 normally has the biggest impact on people compared to the other statements. This is because criticism or acknowledgement at the identity level tends to affect a person significantly, while communication at the environmental level tends to have less impact. Accordingly, when a manager gives feedback to their staff, it is prudent to make comments at the appropriate level; most of the times it is at the behavioural level, thus you would comment on their behaviours. For example, if employees did something good (not extremely great, but good) at the behaviour level (e.g., proofreading a document), and the manager praises them at the identity level (e.g., *'You are great!'*), it does not reflect alignment with the neuro-logical levels. In this case, praise should ideally be given at the behaviour level (e.g., *'Thank you for correcting those errors'*), before (if appropriate) moving up one or two levels (e.g., *'It is great that you can see these details* [capability level]'; *'Details are really important* [beliefs/values level]') (Figure 3.4).

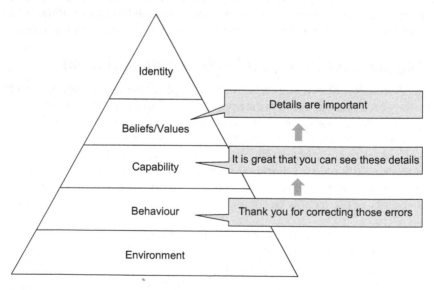

Figure 3.4 The neuro-logical levels in communication.

Example 4: Well-formed outcome

Defining a goal in a way that helps attain it, is important at a workplace. A manager needs to create a personal goal, team goal, and help employees create their own goals. While some frameworks are commonly used (e.g., SMART), NLP provides a comprehensive framework of goal-setting, considering the contexts, five-sensory domains, and required resources, etc. Japanese senior managers who are familiar with a variety of goal-setting methods and frameworks still often find NLP's well-formed outcome to be useful. Here are the eight NLP questions used to help formulate a well-formed outcome (often referred to as the 'eight-frame outcome'; Yamazaki, 2007).

Well-formed outcome (eight-frame outcome)
1. What do you want? (Outcome frame)
2. How do you know when you have achieved it? What would be the visual, auditory, and kinaesthetic information? (Evidence frame)
3. When, where, and with whom would you achieve it? (Context frame)
4. How will other aspects of your life change when you achieve it? (Ecological frame)
5. What resources do you have already that are useful to achieve your goal? (Resource frame)
6. What stops you from achieving it? (Limitation frame)
7. What does achieving it mean to you? (Meta-outcome frame)
8. What is the first action required to achieve it? (Action frame)

It is important here that people state what they *do* want, and not what they do not want. A 'negative goal' refers to a goal that is defined with a negative sentence (i.e., the sentence that includes 'not', 'never', 'no one', etc.), rather than something commonly regarded as negative in society (e.g., sad, depressed). This is an important distinction to make as NLP does not believe in positive and negative emotions *per se*. If your colleague can only think of negative sentences, you can help them to convert these into positive ones using the following process.

1. State what you do not want.—e.g., *I don't want to be a manager like Tom.*
2. Why do you not want it? What are the qualities that make you think so?—e.g., *Because he does not listen to his staff, and takes all the credit of what his team has done.*
3. Which of your personal values do they violate?—e.g., *These violate my values of compassion and respect for my staff.*
4. So would you say that [Value 1], [Value 2] … are important to you (as a manager)?—e.g., *Yes, compassion and respect for my staff are important to me as a manager* (If the answer is no, go back to the question 3, and

redefine the values. You don't have to identify multiple values; often one is enough).
5. Based on those values, what do you want? Or who do you want to be?—e.g., *I want to be a manager who cares and respects my staff.*

The second of the eight well-formed outcome questions explores the evidence frame of the achievement, focussing on the five-sensory information. This is a unique aspect of NLP goal-setting—identifying the visual, auditory, kinaesthetic, olfactory, and gustatory information relevant to attaining the goal. Goal-setting is typically done with words but introducing sensory information can create a physiological difference that maximises motivation. For example, one can think about faces of one's family when the goal is achieved as well as what they might say and the voice tone they will use to say it. The same applies to the third question where clearly identifying the context and environment associated with goal achievement can help to induce physiological and psychological responses that aid goal attainment.

NLP practitioners often find the fourth question unique too, as it can highlight the 'negatives' associated with goal achievement. For example, overly focussing on the professional area of your life might negatively affect other areas of your life. Sometimes the very purpose of working can be sabotaged because the ecology of your goal achievement isn't evaluated. The fifth question explores the resources to hand that can aid goal achievement. This can be an external object and an internal quality, and it can thus include a mentor, family members, facilities, past experience, and qualities of courage and/or resilience, etc. Additionally, one should consider the resources that are lacking (and how they can be obtained) in regard to goal achievement.

The sixth question relates to limitations; what stops you from achieving the goal? It could be lack of experience or ability, but in many cases, limiting self-beliefs are the major obstacle. In this exercise, it is sufficient to just identify one's limiting beliefs, because often they will be eliminated as one takes actions towards goal achievement. We are not going to describe the details here as it's not the main purpose of this chapter, however at this point, you could start the process of belief change, which involves identifying the limiting belief, collecting counter-examples, disputing the limiting belief, creating an alternative belief.

The seventh, meta-outcome is about what happens after achieving the goal. A person may not want to achieve their goal *per se*, but there is some deeper meaning and purpose that they want to realise and satisfy. Finally, the eighth question identifies actions required to achieve your goal. By this point in this exercise, a person should have identified key information to help achieve their goals and should thus be more motivated, wise, and confident in respect of identifying actions.

The well-formed outcome can be used at an individual level and an organisational level. Also, it is useful to review and redefine the goal by frequently revisiting these eight questions. A set of example responses to each of the eight questions is provided below.

Well-formed outcome (eight-frame outcome): example responses

1. What do you want to achieve? (Outcome frame)—e.g., *I want to receive the manager's awards next year.*
2. How will you know when you have achieved it? What would be the visual, auditory, and kinaesthetic information? (Evidence frame)— e.g., *At the next year's award ceremony event in the department, my name is called, and I see my name on the screen. I see my colleagues celebrating it for me, hear their kind words, and feel joy in my chest. I hear me saying in my mind 'I did it!'*
3. When, where, and with whom will you be when you achieve it? (Context frame)—e.g., *Next December, at the ABC conference arena, with my colleagues.*
4. How will other aspects of your life change when you achieve it? (Ecological frame)—e.g., *It is a competitive award, so I may have to work a bit longer on weekdays. But I will make sure to spend quality time with my family on weekends. That will be my priority on weekends. My health needs to be taken care of. I will go to the gym for 30 min three times a week.*
5. What resources do you have already that are useful to achieve your goal? (Resource frame)—e.g., *My family gives me a sense of happiness and safety. I have good colleagues who are willing to contribute to the team's outputs. My company is generally supportive of a good work–life balance. My high work engagement and passion for work are useful to achieve this goal. I am also a good empathetic listener.*
6. What stops you from achieving it? (Limitation frame)—e.g., *I don't have enough experience in designing strategic sales plans, because my main background is in quality management, not in sales. This needs to be learned from training, mentoring, and on-the-job training.*
7. What does it mean to achieve the goal? (Meta-outcome frame)—e.g., *Achieving this goal would give me a sense of self-efficacy that I can do something. I would tell this to my family, and they would be proud of my effort at work. We would probably go on a nice vacation to celebrate this.*
8. What would be the first action? (Action frame)—e.g., *For the most part, I will just keep doing what I have been doing. But as highlighted in this exercise, strategic skills need to be learned. I will send an email to HR and my mentor about this, exploring training or study materials. Also, I will have some time tonight to think about fun things to do with my family on weekends.*

Example 5: Disney strategy

The Disney Strategy was developed by Robert Dilts and is based on an analysis of how Walt Disney achieved his dreams (Dilts, 1998). One salient feature of NLP is modelling through thorough analysis (Bandler & Grinder, 1979). Dilts collected literature about Walt Disney and identified a common pattern in his dream-achieving, which is the development of the Disney Strategy. Japanese managers have reported that this strategy is enjoyable and relatively easily to implement as it has only three positions to be explored. In one intervention study (Kotera & Sheffield, 2017), participants reported that the strategy was particularly useful for them in terms of feeling intrinsic motivation and confidence. The Disney Strategy has also been reported as one of the most useful NLP skills by qualified career consultants in Japan (Kotera, 2018).

The Disney Strategy involves the three perceptual positions: dreamer, realist, and critic (spoiler) positions. Walt Disney often envisioned his dreams with his business partners, one of whom was good at creating realistic plans, and the other was good at finding obstacles in the business' functioning. Each of the three individuals had a different role and function, yet working collaboratively they realised Walt Disney's dreams. In the Disney Strategy, one steps into each of the three positions shown in Figure 3.5.

In the first position of dreamer, the focus is on creating a vision based on what you want to do. The dreamer believes that anything is possible. The body posture of the dreamer is to set your head and eyes upward, and often it is helpful to open your arms and put your palms up. It's a symmetrical and relaxed posture. You think what you want in this position with a belief and attitude of 'anything is possible'. You also want to identify the five-sensory information and context of achieving the dream, as well as the positive impact of doing so. The function of this position is to experience the pleasure of achieving the goal in advance. Remember in this position, you only focus on the dreamer, nothing else (Figure 3.6).

Useful questions to ask in the dreamer position
What do you want to do? (as opposed to what you don't want)
Why do you want to do it?

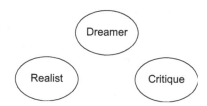

Figure 3.5 Three positions in the Disney Strategy.

Figure 3.6 The Dreamer Posture.

What will you see, hear, and feel when you have attained it?
What will you say to yourself?
When, where, and with whom will you be?
What are the benefits of achieving this?
Where do you want this to take you in the future?
What's the meaning of achieving this dream?
Who do you want to be or be like in relation to manifesting this idea?

Once you have identified your dream, you move to the realist position. Your focus here is on planning—thinking about how (steps) to make the dream come true. The realist acts as if the dream is achievable, and thinks about what steps are required in order to achieve it. In the realist position, you keep your head and eyes straight or slightly forward. Again, it is a symmetrical posture. Clients sometimes prefer to adopt this position while walking on the spot, in order to facilitate their planning. In the realist position, you establish timeframes and milestones for progress. It is helpful to make testable measures to know whether you are moving towards the goal or away from it (Figure 3.7).

Useful questions to ask in the realist position
When will the overall goal be completed?
What will be the 1st, 2nd, and 3rd steps?
Who will be involved? (Assign responsibility and secure commitment from the people who will be carrying out the plan).

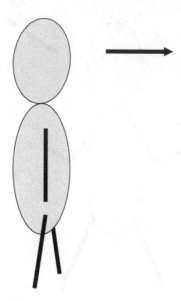

Figure 3.7 The Realist Posture.

What will be your ongoing feedback to show whether you are moving
   towards or away from the goal?
How will you know that the goal is achieved?

Lastly, you step into the critic position, where you review the dream and
plans. Your focus here is on finding what is missing so that you can pre-
vent problems in the future. Your attitude is to consider what to do if
problems occur. As you identify the potential obstacles, if necessary, you
can go back to the realist position to think about how to deal with those
problems. The useful posture to take in this position is an angular posture
with your head and eyes down and tilted. This posture helps you find out
what is missing and needed (Figure 3.8).

Useful questions to ask in the realist position
Is there anything missing in this planning? If so, what can be done?
Will your work–life balance be okay? If not, what can be done?
Will there be anybody who may object to this plan?
What are their needs?
What are the positive gains of the present plan?
Do you think you will need a reviewing or break point? If so, when and
   how will you review/take a break?
What do you need to be careful of, while following through with the
   plan?

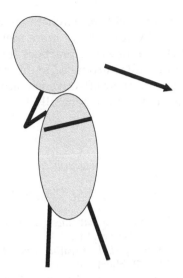

Figure 3.8 The Critic Posture.

Sometimes it is also useful to take a meta-position, which is located outside of the triangle of these three positions. In an intervention study focussing on the Disney Strategy (Kotera & Sheffield, 2017), participants reported that adopting the aforementioned order (i.e., first the dreamer, then the realist, and the critic) was the most useful, and half of the participants felt most familiar with the dreamer position.

The Disney Strategy appears to integrate the eight well-formed outcome questions in one way or another. For example, in the dreamer position, you identify what you want, your sensory experience, context associated with goal attainment, and what happens after achieving the goal. These are similar to the outcome, evidence, context, and meta-outcome frames. The realist explores steps to be taken and resources to be used and needed, which are similar to the action and resource outcomes. The critic position considers what may be missing in the dream and plans, which is similar to the limiting and ecological frames.

Conclusion

In this chapter, we have introduced specific examples of NLP skills used by Japanese senior managers in their day-to-day work. NLP has been utilised in organisational settings in Japan and is widely reported to foster promising results. However, a scarcity of credible empirical evidence for NLP has been criticised by academia for years, yet this does not mean that NLP-active workers should stop doing what works. Indeed, an absence of evidence of effectiveness does not necessarily equate to evidence of ineffectiveness. Furthermore, NLP researchers are currently working to bolster the evidence base by conducting more methodological rigorous NLP studies that can help to refine NLP practices. It is the present authors' hope that the specific NLP skills introduced in this chapter may help individuals seeking to maximise their effectiveness at work, as well as contribute to a healthy life outlook more generally.

References

Bandler, R., & Grinder, J. (1979). *Frogs into princes: Neuro linguistic programming.* Moab, UT: Real People Press.

Bateson, G. (1972). *Steps to an ecology of mind.* Chicago, IL: University of Chicago Press.

Buettner, D. (2017). *The blue zones of happiness: A blueprint for a happier life.* Washington, DC: National Geographic.

Dilts, R. (1996). *Visionary leadership skills: Creating a world to which people want to belong.* Capitola, CA: Meta Publications.

Dilts, R. B. (1998). *Modeling with NLP,* New York, NY: Meta Publications.

HemmatiMaslakpak, M., Farhadi, M., & Fereidoni, J. (2016). The effect of neuro-linguistic programming on occupational stress in critical care nurses. *Iranian Journal of Nursing and Midwifery Research,* 21(1), 38. doi:10.4103/1735-9066.174754

Kaneko, Y., & Sato, I. (2010). *Jisatsu utsu taisakuno keizaitekibenneki [Economic benefits of reducing suicides and depression].* Tokyo: National Institute of Population and Social Security Research.

Kobori, O., Nakazato, M., Yoshinaga, N., Shiraishi, T, Takaoka, K., Nakagawa, A., … Shimizu, E. (2014). Transporting Cognitive Behavioral Therapy (CBT) and the Improving Access to Psychological Therapies (IAPT) project to Japan: Preliminary observations and service evaluation in Chiba. *Journal of Mental Health Training, Education and Practice,* 9(3), 155–166.

Kotera, Y. (2018). A qualitative investigation into the experience of neuro-linguistic programming certification training among Japanese career consultants. *British Journal of Guidance and Counselling,* 46(1), 39–50. doi:10.1080/0306988 5.2017.1320781

Kotera, Y., Gilbert, P., Asano, K., Ishimura, I., & Sheffield, D. (2018). Self-criticism and self-reassurance as mediators between mental health attitudes and symptoms: Attitudes towards mental health problems in Japanese workers. *Asian Journal of Social Psychology.* doi:10.1111/ajsp.12355

Kotera, Y., & Sheffield, D. (2017). Disney strategy for Japanese university students' career guidance: a mixed methods pilot study. *Journal of the National Institute for Career Education and Counselling,* 38, 52–61. doi:10.20856/ jnicec.3808

Kotera, Y., Sheffield, D., & Van Gordon, W. (2018). The applications of neuro-linguistic programming in organisational settings: A systematic review of psychological outcomes. *Human Resource Development Quarterly.* doi:10.1002/ hrdq.21334

Kotera, Y., & Van Gordon, W. (2019). Japanese managers' experiences of Neuro-Linguistic Programming: A qualitative investigation. *Journal of Mental Health Training, Education and Practice.* doi:10.1108/JMHTEP-06-2018-0033

Ministry of Health, Labour and Welfare. (2010). *Shokubani okeru mental health taisakukentoukai houkokusho [Report from the committee for mental health solution at work].* Retrieved from http://www.mhlw.go.jp/stf/houdou/2r9852000000q72m-img/2r9852000000q7tk.pdf

Ministry of Health, Labour and Welfare. (2013). *Roudou anzen eisei tokubetsu chousa (Roudousha kenkou joukyou chousa) no gaikyou [Summary of Labour safety hygiene special research (Workers health status research)].* Retrieved from http://www.mhlw.go.jp/toukei/list/dl/h24-46-50_05.pdf

Ministry of Health, Labour and Welfare. (2015). *Shin job-card seido sokushinkihonkeikaku [Basic plan for promoting the new job-card system].* Retrieved from http://www.mhlw.go.jp/bunya/nouryoku/job_card01/dl/kihonkeikaku.pdf

Ministry of Health, Labour and Welfare. (2016). *Karoushitou boushitaisaku hakusho [Karoshi white paper].* Retrieved from http://www.mhlw.go.jp/wp/hakusyo/karoushi/16/dl/16-1.pdf

National Police Agency. (2016). *Heisei 27nenchuu niokeru jisatsunojoukyou [Suicides in Japan, 2015].* Tokyo: Author.

Organisation for Economic Co-operation and Development. (2015). *Suicide rates (indicator).* doi:10.1787/a82f3459-en (Accessed on 31 May 2015).

Rao, D. H., & Kulkarni, D. G. (2010). *NLP for stress mitigation in employees.* Paper presented at International Conference on Education and Management Technology (ICEMT). doi:10.1109/ICEMT.2010.5657585

Yamazaki, H. (2007). *NLP no kihon ga wakaruhon [Understanding the basic of NLP].* Tokyo: Japan Management Center.

chapter four

Neurolinguistic programming for managers and engineers as evidence-based practitioners

Tomasz Witkowski
Polish Skeptics Club

Stawomir Jarmuz
Moderator Ltd.

Contents

Introduction: Managers and engineers as evidence-based practitioners

When a crime is committed, the police gather and secure evidence. If we were to appear in court, standing accused of that crime, and we were not presented with sufficient proof, we would likely conclude that our rights had been violated. We rely on evidence when we sit down on a roller coaster. We trust that the engineers gathered enough evidence on the durability of the particular materials used in its construction. We also trust that the doctor treating us is applying evidence-based methods, and if this were not the case, we'd probably get quite angry with him/her. Why is it that in so many situations we seek out evidence and rely on it?

Likely because evidence-based actions deliver good results. Evidence-based medicine led to a significant increase in humans' average lifespan in the twentieth century, and diseases which were fatal just a few decades earlier are today a far lesser threat. The development of technology we are both witnesses to and beneficiaries of is possible only because of serious consideration being given to evidence. On the other hand, actions undertaken without evidence or in which evidence is given insufficient consideration frequently end in disaster, or even tragedy. Are we, however, rational enough to seek out in every sphere of activity evidence that would serve to guarantee our effectiveness? Do we ask the teachers teaching our children for evidence? Are their methods based on evidence? How much weight do managers responsible for managing people give to evidence? Or trainers in charge of training our employees? As it is, the answer here is far less than clear. Many methods of teaching or management are based on evidence, but many of them are simply "common wisdom," or long-standing myths that can do more harm than good. Why, then, do we depend solely on evidence in some areas, while in others we do not?

We can identify two main causes of this phenomenon:

The first—the application of ineffective methods in teaching or management rarely leads to spectacular catastrophes. The negative effects consist in insignificant wasting of time at trainings, but frequently for a large number of people. This is also an inefficient use of financial resources, but almost never of such sums that would threaten the company.

The second cause results directly from the first. Because the failures that derive from the application of ineffective methods are never spectacular, we have become accustomed to adopting a very tolerant approach to evidence, and we frequently accept subjective experiences as evidence. We demand that an engineer present us with hard calculations, but from training methods we often only require that they be attractive and appealing to their participants. Is this a rational approach? Shouldn't we ask about evidence with greater frequency? Shouldn't we look for the most effective methods? The question is, however, how should we do so?

Most importantly, we should know that almost anything can constitute evidence. The testimony of an eyewitness who has seen a UFO can be evidence confirming the existence of extraterrestrials. In turn, a picture of someone wandering around the mountains can become evidence of the existence of beings referred to as Yeti. The recommendations of patients can be evidence of the effectiveness of a doctor, while the value of training and developmental methods can be attested to in the opinions of participants. We can identify many such kinds of evidence, but scientists do not treat them as decisive; indeed, the essence of evidence-based practices is the search for the best available evidence, and not just any evidence at all. Here, however, another question arises that may seem a difficult one at first glance: how can we determine whether a given piece of evidence is of greater or lesser value? How to orient oneself in the thicket of methods and means of distinguishing which piece of evidence is worthy of trust? Thankfully, we are aided by scientists, who have wondered for decades about which proofs should be given the greatest weight, and which should be approached with caution. We can be aided in this respect by the broadly respected hierarchy of evidence (e.g., Ng & Benedetto, 2016; Page & Meerabeau, 2004; Petrisor & Bhandari, 2007). Despite minor differences among authors, the majority of them admit that the evidence we cite is at the bottom of the hierarchy. The majority of scientists also agree with the statement that the highest level is occupied by umbrella reviews, meta-analyses, and systematic reviews. One rung lower are randomised controlled trials, and lower still are quasi-randomised trials and observational studies.

Desiring to act rationally, whether in making a decision about a training seminar, or in selecting a management method, we should always examine on what level of the hierarchy of evidence are the proofs we are using to justify our actions. Basing one's actions on evidence lower in the hierarchy when there is evidence of greater significance available to us brings to mind an old Indian tale of blind men who decide to learn about an elephant. One of them, upon touching the elephant's hard skin, said it was a wall. The second touched a tusk, and said it was a spear. The third, having touched the elephant's trunk, was convinced he had encountered a sort of snake. The fourth, brushing up against the elephant's knee, said it was a tree. Having felt a gust of wind caused by the elephant's breathing, the fifth blind man called it a fan. And, finally, the sixth, having grabbed the elephant by its tail, said that it was just a normal piece of rope.

The evidence-based approach was developed in places where opinions and viewpoints can have no truck—in saving human lives, in the work of engineers. It has proven itself, which is why thinking managers employ it in managing, good teachers in teaching, and good trainers in trainings. There is also increasingly frequent mention of evidence-based politics. We recommend this approach to all practitioners, which is why we have performed an analysis of the available evidence regarding the effectiveness of neurolinguistic programming that can serve as an aid in taking decisions about using it for one's own development, as well as for needs related to employee training.

The outline of the model (as created by Bandler and Grinder)

In the 1970s, Richard W. Bandler and John Grinder had the idea to create a practical therapy model. They reasoned that a group of recognised psychotherapists acted on the basis of implicit theories that helped them achieve substantial psychotherapeutic effectiveness and great rapport with clients. Grinder and Bandler concluded that careful observation of these skillful therapists at work should lead to the identification of successful patterns of practice that would then be empirically verified and disseminated to other practitioners. For several years, they observed such therapists as Fritz Perls, Milton H. Erickson, and Virginia Satir at work. Based on these observations and reflections, Grinder and Bandler formulated neuro-linguistics programme's (NLP's) tenets and hypotheses. With time, their strategy was promoted as a "science of excellence." It reflected a procedure also known as "modelling," in which one studies the performance of highly successful people from different walks of life in order to learn skills to improve one's own personal and professional life (O'Connor & Seymour, 1993).

The originators of NLP described it as a "model" rather than a "theory." The central philosophy of the NLP model is summed up in the phrase "the map is not the territory" (e.g., Lankton, 1980, p. 7). The idea is that each individual operates on the basis of his or her internal representation of the world (the "map") and not the world itself (the "territory"). One's interactions with the world are formed by mental maps created from one's surroundings. The maps are by nature distorted, limited, and inflexible. The task of the therapist is to understand the client's particular map and to convey that understanding to the client. As Heap (2008) and Newbrook (2008) maintained, this philosophy was verbalised by the philosopher and linguist Korzybski (1933).

At the core of NLP lies the notion of a preferred representational system (PRS). It is argued that the maps people make of their world are represented by five senses: visual, kinaesthetic (referring to both tactical and visceral sensations), auditory, olfactory, and gustatory. Every experience is composed of information received through these sensory systems. NLP proponents coined the term "representational systems" to describe the patterns in the way that sensory data are represented in people's minds (Bandler & Grinder, 1979; Grinder & Bandler, 1976). They claimed that every person processes most information using predominantly one PRS.

Another ostensible discovery of which NLP originators were particularly proud was the idea that access to the representational systems is possible through so-called accessing cues, and more specifically, through eye movements (EMs). They explained, for instance, that a person engaged in a cognitive activity in the visual mode would tend to look upwards, whereas a person using the auditory model would tend to look horizontally (Bandler & Grinder, 1979). The kinaesthetic mode is believed to be associated with a gaze downwards to the right. Thus, careful observation of such EMs should enable an NLP therapist to unequivocally identify the PRS of a client or an interlocutor and, as a consequence, facilitate a therapeutic intervention matched to the individual's particular PRS.

The final assertion is that by matching, mirroring, or pacing clients' verbal and non-verbal behaviour, including their PRS, the NLP practitioner has the opportunity to achieve effective communication, gain the client's trust, and enhance rapport. Reflecting on the example of model therapists, in order to work effectively with a client, the therapist should strive to match the client's PRS to be able to use his or her "map." A trained practitioner can identify the method in which information is stored through careful observation of the client's eye-gaze patterns, posture, breathing patterns, tone of voice, and language patterns. Subsequently, the therapist takes every care to match the clients' PRS while communicating with them. PRS matching effected through following the same language patterns—i.e., by using predicates typical of the mode operated by the client—ostensibly yields the best results in facilitating communication and enhancing therapeutic effectiveness.

The NLP originators promoted it as a sensationally effective and rapid form of psychological therapy. They claimed that NLP helped overcome such problems as phobias and learning disabilities in less than an hour's session, whereas with other therapies, progress might take weeks or months (e.g., Bandler & Grinder, 1979, p. ii; Lankton, 1980, pp. 9–13). They also maintained that a single session of NLP combined with hypnosis could eliminate certain eyesight problems, such as myopia (Grinder & Bandler, 1981, p. 166), and even cure the common cold (Grinder & Bandler, 1981, p. 174). Today, we can find similarly alluring promises in many NLP programmes and advertisements.

The development of NLP can be analysed in terms of a particular research paradigm within social psychology known as the "full cycle" approach, developed by Cialdini (1980). Cialdini critically argued that experimental lab-based research, although offering the opportunity to control variables and establish causality, had several accompanying weaknesses. The most prominent was the inability to determine the strength or prevalence of phenomena in the natural environment. Cialdini argued instead that researchers should use naturalistic observation to determine the presence of an effect in the real world, then develop a theory to determine what processes may underlie the effect, followed by experimentation to verify the effect and its underlying processes, and finally return to the natural environment to corroborate the experimental findings.

Bandler and Grinder omitted the critical stage of empirical verification of their assertions. They found that part of the process redundant and unnecessary, and so moved directly to the formulation of the model and to putting it into practice. They were known for their openly demonstrated contempt for the scientific verification of NLP hypotheses:

> As far as I can tell, there is no research to substantiate the idea that there is eyedness. You won't find any research that is going to hold up. Even if there were, I still don't know how it would be relevant to the process of interpersonal communication, so to me it's not a very interesting question. (Bandler & Grinder, 1979, p. 31)

Nevertheless, many subsequent researchers did subject NLP to empirical evaluation.

Review of research to date

Beginning in the 1980s, a wide range of articles on NLP were published. Particular attention should be paid to several reviews of research on the effectiveness of NLP. The first two were conducted by Sharpley (1984, 1987). In his first analysis, he reviewed 15 studies examining the possibility of identifying and matching clients' PRS. Sharpley (1984) concluded that "the identification of this PRS (if it is a PRS and not merely current language style) by either eye movements or self-report is not supported by the research data ... The existence or stability of the PRS is irrelevant to predicate matching as a counselling process, and parsimony argues for the process rather than the yet unverified theory ... Of most importance, there are no data reported to date to show that NLP can help clients change" (p. 247).

The second review (Sharpley, 1987) is even more conclusive. It was written as a response to a critical paper by Einspruch and Forman (1988), who analysed 39 studies on NLP and described methodological errors and a lack of sufficient knowledge about the theoretical underpinnings of NLP demonstrated by authors. Sharpley took into account works investigated by Einspruch and Forman and expanded that sample with additional ones to perform an analysis of 44 studies. Six papers (13.6%) provided evidence supportive of NLP-derived theses, 27 (61.4%) failed to lend support for one or more of those tenets, and 11 (25%) showed only partial support. Sharpley examined all works available, starting from doctoral dissertations to papers published in high-ranking peer-reviewed journals. He summed up his review as follows:

> There are conclusive data from research on NLP, and the conclusion is that the principles and procedures suggested by NLP have failed to be supported

> by those data ... Certainly research data do not sup-
> port the rather extreme claims that proponents of
> NLP have made as to the validity of its principles or
> the novelty of its procedures. (pp. 105–106)

There are also other important research reviews. In 1988, Heap analysed 63 studies and concluded that the assertions of NLP writers concerning representational systems had been objectively investigated and found to be wanting. The hypothesis that it is possible to identify PRS through careful observation of EMs was likewise not confirmed. In Heap's view, the stated conclusions and the failure of investigators to demonstrate convincingly the alleged benefits of predicate matching seriously questioned the role of such procedures in counselling. Dorn, Brunson, Bradfor, and Atwater (1983) similarly concluded from their review that there was no demonstrably reliable method of assessing the hypothesised PRS.

If the NLP assertion of a reliably identifiable PRS and the corresponding ability to enhance communication through PRS matching proved to be true, it would have important implications for neuroscience, cognitive psychology, and a number of other disciplines. If the NLP claims concerning its instant effectiveness proved to be true, the field of psychotherapy would be transformed, perhaps even revolutionised, and research reporting the effectiveness of therapy would position NLP as a potential first-rate therapy. Nothing of the sort has taken place. Instead, experts warn against using NLP, and classify it is as one of many dubious "fringe therapies" (Beyerstein, 2001) or "power therapies" (Devilly, 2005). NLP is also found on many lists of discredited therapies. Norcross, Koocher, and Garofalo (2006) sought to establish consensus on discredited psychological treatments and assessments using the Delphi methodology. A panel of 101 experts participated in a two-stage survey, wherein they had to report familiarity with 59 treatments and 30 assessment techniques and rate them on a continuum from *not at all discredited* (1) to *certainly discredited* (5). Neurolinguistic Programming for treatment of mental/behavioural disorders was assessed at an average of 3.87 (SD = 0.92). For comparison, "angel therapy for treatment of mental/behavioral disorders" obtained the highest score, $M = 4.98$ (SD = 0.14), and "behavior therapy for sex offenders" obtained the lowest, $M = 2.05$ (SD = 0.91).

Similarly, Roderique-Davies (2009) cautioned against applying NLP to therapeutic practice in an article *Neuro-linguistic programming: Cargo cult psychology?* in which he concluded:

> NLP masquerades as a legitimate form of psycho-
> therapy, makes unsubstantiated claims about how
> humans think and behave, purports to encourage
> research in a vain attempt to gain credibility, yet
> fails to provide evidence that it actually works. (p. 62)

The Army Research Institute (ARI) asked the National Research Council to assess the credibility of "New Age" techniques, including NLP, considered for implementation by the US Army with a view to enhancing human performance. The committee in charge of the review also challenged the effectiveness of NLP. They concluded:

> The conclusion was that little if any evidence exists either to support NLP's assumptions or to indicate that it is effective as a strategy for social influence. (Swets & Bjork, 1990, p. 90)

Von Bergen, Soper, Rosenthal, and Wilkinson (1997) analysed areas of application of "alternative training techniques" in human resources management, and likewise concluded:

> We identified four alternative training techniques that have been widely touted and sold to government and industry: subliminal stimulation, mental practice, meditation, and NLP. Finding that the claims made for three of these techniques, mental practice being the exception, did not stand up to scientific scrutiny, we encourage HRD professionals to guard against substituting testimonials and popularity in the marketplace for research evidence when they consider a new training aid. (p. 291)

An evaluation of the effectiveness of NLP as a communication theory and its contribution to clinical practice and family therapy is similarly unflattering:

> The academic verdict on NLP is given: NLP's theory of the person cannot account for the wide range of intrapsychic and interpersonal problems encountered in clinical practice. A final verdict is withheld until further clinical studies and experimental investigations are reported. (Baddeley, 1989, p. 73)

In 2010, another review of research on NLP was published (Witkowski, 2010). It presented the concept of NLP in the light of empirical research in the Neuro-Linguistic Programming Research Data Base. From among 315 articles, the author selected 63 studies published in journals from the Master Journal List of ISI. Out of 33 empirical studies, 18.2% show results supporting the tenets of NLP, 54.5% results non-supportive of the NLP tenets, and 27.3% report uncertain results. The qualitative analysis indicates the greater weight of the non-supportive studies and their greater methodological worth against the ones supporting the tenets. Analysis results contradict the claim of empirical bases of NLP.

Two years later, a systematic research review was published (Witkowski, 2012). In order to examine the scientific status of NLP, 401 publications were identified, including a subsample (n = 66) that was published in prestigious journals. Of the entire sample, 21 methodologically sound empirical studies published since the last comprehensive reviews of NLP were identified (Heap, 1988; Sharpley, 1984, 1987). Of these studies, 9.5% were found to be generally supportive of the tenets of NLP and/or the effectiveness of NLP techniques, 19% were partially supportive, and 71.5% were non-supportive. The author concluded his review:

> NLP is ineffective both as a model explaining human cognition and communication, and as a set of techniques of influence and persuasion. (p. 37)

Also in 2012, Sturt with her co-workers published a systematic review of the effects of NLP on health outcomes. To evaluate these effects the researchers searched: MEDLINE®, PsycINFO, ASSIA, AMED, CINAHL®, Web of Knowledge, CENTRAL, NLP specialist databases, reference lists, review articles, and NLP professional associations, training providers, and research groups. Their searches revealed 1,459 titles from which ten experimental studies were included into the review. Five studies were randomised controlled trials (RCTs), and five were pre-post studies. Targeted health conditions were anxiety disorders, weight maintenance, morning sickness, substance misuse, and claustrophobia during MRI scanning. NLP interventions were mainly delivered across 4–20 sessions, although three were single-session. Eighteen outcomes were reported, and the RCT sample sizes ranged from 22 to 106. Four RCTs reported no significant between-group differences, with the fifth finding in favour of the NLP arm. Three RCTs and five pre-post studies reported within-group improvements. The authors concluded that there is little evidence that NLP interventions improve health-related outcomes. In their opinion, there was insufficient evidence to support the allocation of National Health Service (NHS) resources to NLP activities outside of research purposes (Sturt et al., 2012).

The only meta-analysis on NLP was published in 2015 (Zaharia, Reiner, & Schütz, 2015). The authors removed 350 from a total number of 425 studies, considering them not relevant based on the title and abstract. In the final analysis, they included 12 studies, with numbers of participants ranging between 12 and 115. The vast majority of studies were prospective-observational. The article evaluates the effectiveness of NLP therapy for individuals with social/psychological problems. The overall meta-analysis found that NLP therapy may add an overall standardised mean difference of 0.54, with a confidence interval of CI = [0.20; 0.88]. In the opinion of the authors, NLP as a psychotherapeutic modality shows results that can hold their ground in comparison with other psychotherapeutic methods.

As the results of this meta-analysis run contrary to the conclusions derived from former reviews, they must be taken into account very carefully. First of all, it is not quite clear how the researchers reduced the number of studies from 75 to 12. Although they write about excluding criteria such as "Not the right population; studies conducted on healthy individuals with social/psychological problems (n = 19); Not the right intervention (n = 17): studies conducted in healthy individuals with social/psychological problems (n = 8), depression (n = 5), other (n = 4); Not the good outcome: studies carried out in healthy individuals with social/psychological problems (n = 17); Excluded based on study design (n = 11): review, editorial, comment letter, study design protocol. (p. 357)," we are unable to see what studies were excluded. Moreover, the authors included into their sample also unpublished and never peer-reviewed data. Also of significance is that all authors of this article are active practitioners of NLP, which may give rise to conflicts of interest.

Today, after 40 years of research devoted to the concept, NLP is closer to the image of an unstable house built on sand rather than an edifice founded on empirically-based rock. In 1988, Heap passed verdict on NLP. As the title of his article indicated, this was an interim verdict. In his conclusions, he wrote:

> If it turns out to be the case that these therapeutic procedures are indeed as rapid and powerful as is claimed, no one will rejoice more than the present author. If however these claims fare no better than the ones already investigated then the final verdict on NLP will be a harsh one indeed. (p. 276)

We are fully convinced that we have gathered enough evidence to announce this harsh verdict now, but before we do so, we would like to take a thorough look at the practice of NLP.

Between theory and practice—a review of the present content of NLP training methods

NLP is a collection of diverse methods and techniques that would be hard to call an obligatory canon. NLP practitioners in different countries use chosen methods in combination with their own inventions. We decided to describe and assess a few of the methods most frequently appearing in books on NLP and in training programmes. The selection is thus necessarily a subjective one, but it encompasses the most important offerings one can come across at NLP courses, in books, and in internet-based materials.

Presuppositions

The definition of "presupposition" is an assumption, a judgement, a condition. More precisely, an assumption we silently adopt in many of our statements. The term comes from linguistics, with a tradition in science reaching back to the mid-nineteenth century. Considerations addressing similar issues, however, can be traced back to the work of Aristotle, and are present throughout the entire history of philosophy (Horn, 1996). Let us recall here that NLP, which generally does not address the roots of presupposition, came about in the 1970s. Thus, existing knowledge was incorporated into it, and the "user" of NLP is generally unaware of its true roots.

The majority of our utterances contain some assumptions. Without them, it is impossible to say whether a sentence is true or false in the classic sense of truth as being in accordance with reality. In the sentence "John likes trainings about emotional intelligence for managers," we are faced with several presuppositions. First, that there really is a person named John. Second, that there are trainings for managers on the subject of emotional intelligence. And third, that John participates in them. In the absence of these assumptions, we would be unable to declare whether the sentence is true or false. These kinds of presuppositions seem obvious. However, presuppositions related to attempts at persuasion are more interesting. The sentence "When will you decide to get a new job?" contains the hidden assumption that the addressee is considering changing employment. Perhaps this individual has not yet in fact taken a decision, and the utterance serves to induce such a step. In the context of negotiations and sales, a sentence with a presupposition could go like this: "Which model do you choose, the first or the second?" The client is pressed into making a choice, while in fact no decision has yet been made by that person as to whether or not to buy. Presuppositions of this nature open up space for social influence, and often—manipulation. And this is precisely why they have been incorporated into NLP as verbal devices for achieving goals. The tools of linguistics are employed in such fields as politics, advertising, and marketing. NLP in these cases is not creating theory, but rather taking advantage of scientific achievements under its own banner. This is analogous to developing a new field of knowledge with a fancy-sounding name, while simply teaching arithmetic under its aegis. The Middle Ages philosopher William Ockham developed the famous rule which bears his name, "Ockham's Razor," also known as economy of thought. This rule says that when seeking to explain some phenomenon, the number of notions employed should be limited. NLP violates the precept of "Entia non sunt multiplicanda praeter necessitatem."

Metamodel

The metamodel—like presuppositions—is a linguistic device. Some NLP proponents treat presuppositions as a part of a metamodel. Here, however, we treat these two issues separately. The metamodel serves the analysis of language and uncovering of hidden meanings contained in a message, which serves to facilitate communication. The conception was elaborated by the creators of NLP, Richard Bandler and John Grinder, in the first book written on the subject (Grinder & Bandler, 1976). In their work, the progenitors of NLP adopt the following assumption:

> Language serves as a representational system for our experiences. Our possible experiences as humans are tremendously rich and complex. If language is adequately to fulfil its function as a representational system, it must itself provide a rich and complex set of expressions to represent our possible experiences. (p. 24)

According to Grinder and Bandler, the problem is that language frequently deforms real experience. This is why tools are needed for messages to refer to the greatest degree possible to experiences and facts. The metamodel distinguishes three processes that alter the precise meaning of a message: generalisation, deletion, and distortion. **Generalisation** consists in drawing general conclusions from individual events; for example, when an employee does not perform his task correctly, the manager says: "He never does his work properly." This generalisation disfigures the image of the employee if he has performed other tasks correctly. Generalisations are characterised by the application of general quantifiers, e.g., such words as every, everybody, always, never, etc. Another example of generalisation comes in the form of imprecise verbs, whose general meanings can be understood in different ways. A typical example in business relations is the assurance "I will do everything in my power." We don't know what, specifically, that person will do in that particular situation.

The second process altering the meaning of a message is **deletion**. It occurs when certain aspects of information are omitted, which leads to an incorrect conclusion. In a business situation, the salesperson says to the client: "Our solution is the best." We don't know exactly what he/she has in mind, nor with what other solutions he/she is comparing his/her offer to. The deletion of this information leads to a warping of the general assessment. The third process distinguished in the metamodel is **distortion**, which consists in employing notions or terms that can be understood differently in different contexts. Grinder and Bandler argue that distortions are not only a linguistic process, but also constitute the basis of artistic activity. Creation consists in a modification of reality. There are many instances of distortions described by NLP practitioners. One of them is "mind reading," consisting in assigning a specific intention to someone, for example "There's no way that you like me." This kind of message makes it difficult to maintain good relations. Another example is the lost performative, which consists in creating sentences without subjects, e.g., "That system shouldn't be activated." The absence of a subject makes the message sound like an objective need or necessity, and makes discussion of a problem difficult.

The metamodel is not limited to detailing processes that make it harder to drill down to real meanings, but it also offers tools to enhance communication. These are questions uncovering real meaning or intentions. Here are some examples that relate to the ones presented in the preceding passages:

GENERALISATION

"He never does his work properly." "Never? So what would you say about the project he oversaw that got rave reviews?"

"I will do everything in my power." "What, exactly, will you do? What will the result of your work be?"

DELETION

"Our solution is the best." "In what sense? Compared to what solutions?"

DISTORTION

"There's no way that you like me." "For what reasons do you think so? And what could convince you that I in fact *do* like you?"

"That system shouldn't be activated." "Who thinks we *should* activate that system? What are the arguments in favor of activating it? What will happen if we don't?"

Initially, the metamodel was used by NLP practitioners in therapy, and then it entered the world of trainings and personal development. As opposed to many of the offerings of NLP, the metamodel is a useful tool that raises no doubts, and which primarily serves to improve communication in diverse social contexts. Because the processes of generalisation, deletion, and distortion at times are put into service of the goal of manipulation, the questions proposed in the metamodel are a good way of coping with manipulation. The usefulness of this area of NLP is likely linked to strong inspiration from psycholinguistics, and thus a field of science with a strong empirical foundation.

Anchoring

Another tool we can identify in the programmes of the majority of NLP courses is anchoring. Because the NLP literature contains various definitions and descriptions of this device, we have selected that which is presented on Wikipedia. It links elements of several others proposed by leading proponents of this trend.

> NLP teaches that we constantly make *anchors* (Classical Conditioning) between what we see, hear and feel and our emotional states. While in an emotional state if a person is exposed to a unique stimulus (sight, sound or touch) then a connection is made between the emotion and the unique stimulus. If the unique stimulus occurs again, the emotional state will then be triggered. NLP teaches that anchors (such as a particular touch associated with a memory or state) can be deliberately created and triggered to help people access 'resourceful' or other target states. (Methods of neuro-linguistic programming, n.d.)

In using anchoring, NLP practitioners first define what emotions we will require in a given situation. These can be courage, serenity, joy, as well as others. In the second step, we should recall a situation in our life when we had a clear experience of a given emotion, e.g., when listening in our youth to a particular piece of music, we felt joy. The following step is linking that recalled situation with some sort of neutral gesture, such as folding our arms, an image, or some other stimulus referred to as an anchor. According to NLP practitioners, the association of a neutral gesture with a recalled situation creates conditioning that evokes the pleasant, desired emotion. In the opinion of NLP proponents, in this manner we can cope with negative emotions by substituting positive ones in their place. If we experience fear before a public appearance, it is enough to perform the appropriate gesture that will evoke positive mental associations, including the desired emotion. If this emotion is courage, then the fear is conquered, driven away by positive affect. This process should be repeated in order to make the association permanent.

The anchoring technique seems a simple and useful one. However, from the scientific point of view, it is average at best. Apart from theoretical analysis, this is attested to by empirical study. Several researchers have compared the usefulness of this technique with traditional methods of coping with public appearances. As it occurred, contrary to the promises of NLP's creators describing the negation of fear before public appearances in the course of one session, anchoring generated effects no better than those of other methods (Krugmanet al., 1985). The explanation of the ineffectiveness of this device is associated with a failure to understand the mechanism of classical conditioning that some NLP practitioners systematically invoke.

The mechanism of classical conditioning was discovered and described by the Russian physiologist, Ivan Pavlov. It consists in an unconditional stimulus, such as the appearance of food, evoking an unconditional reaction, i.e., drooling. If the unconditional stimulus appears concurrently with some other stimulus referred to as conditional—for example, the flashing of a lamp—then after several repetitions of the sequence, the conditional stimulus will itself evoke the unconditional response. Conditioning can also apply to negative situations. Explanations of this mechanism in psychology often invoke a study during which a young boy named Albert was playing with his favourite white rat. During the child's play, an experimenter placed behind a curtain hit a metal bar, evoking fear in the child. After a few repetitions, the child began crying on sight of the rat, and generalised this reaction to other white animals. The described mechanism was discovered in the first half of the twentieth century, and constitutes an important contribution to knowledge on the subject of learning by animals, including by humans (Moore, Manning, & Smith, 1978). Let us observe that the link between the unconditional stimulus and the conditional relation relates to activities important to survival, and requires a real, rather than imagined concurrence. Such unconditional reactions as drooling, fear, or sexual arousal occur as an effect of real stimuli (food, loud noise, object of sexual desire). Merely imagining them does not evoke emotion, or at least not strong emotion. If it were possible through force of imagination to create any and all emotions, people would probably take advantage of that possibility continually, without a second thought. NLP suggests the possibility of evoking emotions by way of imagination. What is more, these emotions are linked with a neutral gesture, such as folding one's arms, treated as a conditional stimulus. Since there is a very slight link between imagination and emotion, there should be a slighter a link between the trivial action of folding one's arms and emotions. We add that classical conditioning concerns primary emotions. When it comes to complex emotions—which, in the opinion of NLP proponents, are to be evoked by anchors—the mechanisms are more complicated. We are thus dealing with an entirely providential interpretation of classical conditioning as justification for the effectiveness of anchoring.

Even more confusing is the application of this technique in real situations. Referring to the problem with public appearances, NLP practitioners state that negative feelings, such as nervousness or fear, can be substituted with positive emotions such as confidence or joy. It's enough to simply apply anchoring. Unfortunately, things are not so simple. In psychology, there are very thorough examinations and descriptions of the effect known as "bad is stronger than good" (Baumeister, Bratslavsky,

Finkenauer, & Vohs, 2001). It turns out that negative events or emotions have a much stronger influence on us than positive ones. This is linked with the evolutionary history of our species, when reacting to negative or threatening stimuli was far more important for survival than reacting to pleasant situations. This is why, for example, we have a far stronger sense of smell when it comes to unpleasant odors as compared to pleasant ones. Turning back to the advice of NLP practitioners, in difficult situations such as public appearances, anchoring is supposed to replace negative emotions with positive ones. Taking into consideration the phenomenon of "bad is stronger than good," negative emotions associated with that situation will likely be far stronger than those positive ones that are (if at all) evoked in the imagination. If anchors are to work at all, then, paradoxically, they are far more effective in evoking negative emotions. Thus, on grounds of established psychological knowledge, the practice of anchoring would seem an ineffective one, and the absence of studies confirming the effectiveness of anchoring would seem to reinforce this conviction.

Metaprograms

In NLP training programmes, we may frequently encounter so-called metaprograms, that is, methods people employ for processing information and creating action plans based on that information. According to Hoag (2018), "metaprograms are mental processes which manage, guide and direct other mental processes." The conception of metaprograms assumes that people unconsciously filter information in their own characteristic manner, which in consequence modifies their behaviour. Thus, recognition of a person's metaprograms allows us to predict his/her behaviour, to improve communication, and to exert influence.

This general description is best illustrated using examples. One of the primary metaprograms is *Toward vs Away From*. An individual possessing the metaprogram *Away From* (more specifically, a plan resulting from the metaprogram) focuses on problems, predicts difficulties, and has difficulty in setting goals. He/she frequently uses phrases like "I won't," "I don't want to," "It can't be done," etc. Such people are more easily manipulated, because they are more concerned with duties rather than personal aspirations. In turn, people with the metaprogram *Toward* work with constancy to achieve set goals. Their thinking is so strongly oriented towards achieving results that they fail to perceive barriers. Their vocabulary frequently features such words as "I want," "I'll get it," and "I will achieve". It is more difficult to influence people with the *Toward* metaprogram than *Away From*.

Another metaprogram, *Self-Reference vs Other Reference*, concerns focusing attention on oneself or on others. In the first case, assessment of events is associated with personal observations, experiences, or values, and most frequently serves to maintain one's high feelings of self-worth.

If, for example, someone points out a mistake made by that person, the reaction comes in the form of a statement such as "That's your opinion." When we ask how they know they've done something well, they say "That's how it seems to me." The opposite tendency is demonstrated by people on the other side of the *Self-Reference vs Other Reference* metaprogram. The criterion for information intake is other people, their needs, and values. When pointing out to such people that they have erred, we hear that they need to consult with others and take corrective action. In turn, when asking how they recognise that they have done their work well, they invoke the opinions of others. According to NLP practitioners, the *Self-Reference vs Other Reference* metaprogram is not equivalent to the well-known extroversion–introversion dimension. Initially, eight metaprograms were detailed within NLP, and more were successively added. In 2009, there were around 60 of them (Vaknin, 2010).

This short description of two metaprograms shows that they were intended to be based on a cognitive approach to personality. The essence of this approach is the statement that the manner in which a person takes in and processes information impacts emotions and behaviour. The cognitive approach is very important in contemporary psychology. Although the conception of metaprograms refers in its assumptions to mainstream psychology, it has some very serious flaws. First, metaprograms divide people into a dichotomy: those who prefer one of two opposing scenarios. From the perspective of present-day knowledge gathered in the field of social psychology, the situation has a very strong impact on interpretations of events and on behaviour. This does not mean that internal factors such as personality play no role. However, in teaching managers to understand their own behaviours and those of others through the framework of metaprograms, we frequently play loose and fast with the truth. Second, the conception of metaprograms attempts to describe differences between people, but does so inaccurately, or at least there is no empirical data indicating the sufficiency of metaprograms.

In the PsychINFO database of refereed articles (August 2018), a search for the term "metaprograms" returns only one empirical study, and that is a qualitative one (Brown, 2004). This attests to the lack of interest by scientists in verifying this conception, or to its theoretical opacity. This situation could result from the fact that metaprograms do not form a comprehensive conception of personality, cognitive styles, or other individual differences. They are a collection of ideas inspired in part by practice, and in part by the intuition of NLP trainers. In addition, in NLP publications and programmes there are further metaprograms on a different level of generality, with the relations between particular constructions unclear. The myriad of metaprograms leads to problems with explaining and predicting behaviour, because it is difficult to classify a specific behaviour to one metaprogram. What is more, distinguishing and applying several dozen metaprograms becomes practically worthless. The average person is not capable of analysing others' behaviour applying dozens of explanatory categories.

In summary, metaprograms are an attempt at describing individual differences between people and making practical use of these differences. Studies have not proven the accuracy of this conception. Thus, in spite of its cognitive foundations, it is theoretically immature. From the practical standpoint, the conception of metaprograms is difficult in application.

The Dilts Pyramid

Robert Dilts is a leading NLP authority. He has written a number of books forming the canon of this field (e.g., Dilts, 1983, 1998). The most frequently applied in NLP is the conception known as the Dilts Pyramid, or the Dilts Logical Pyramid, and also Dilts' NeuroLogical Levels. According to Dilts, who drew inspiration from the anthropologist Bateson (1979), the individual should be examined on several levels. Each of these levels is a system in and of itself, and influences the remaining ones. At the bottom of the pyramid is *environment*, i.e., the conditions in which the person lives: social surroundings, physical environment, weather, food, sounds, etc. The environment is perceived by our senses, and this perception is formed by our nervous system. In Dilts' opinion, we can begin to understand a person starting from the way she perceives and analyses her environment.

The second level is *behaviour*, that is, reactions and actions that a person undertakes in his environment. The psychomotor system is responsible for this level.

The next, third level, is our *capabilities*. Dilts understands capabilities as a "mental map" that organises experience. Without capacities, our behaviours would be only a response to stimuli, a collection of habits. It is capabilities as an internal structure that is responsible for a given set of behaviours demonstrated by an individual. Capabilities make it possible to acquire proficiency or mastery in a given class of behaviours. The physiological base of this level is the activity of the "grey matter" of the cortex, where information from the senses is organised in the form of mental maps.

A higher level of the Dilts Pyramid concerns *beliefs and values*, which refer to fundamental judgements about the world, other people, and oneself. Beliefs and values determine the meaning we assign to our surroundings and experiences. They are also the foundation of motivation to action. The physiological base of beliefs and values is comprised of the limbic system and hypothalamus. As Dilts claims, although these parts are more primitive than the cortex, they integrate information from the brain complex and regulate the activity of the autonomous nervous system, such as the heart action.

Going even higher in the Dilts Pyramid, we reach the level of *identity*, that is, the answer to the question "Who am I?" The feeling of identity impacts beliefs and values, establishes missions, and determines goals and roles performed. As opposed to the lower levels for which particular parts of the human nervous system are responsible, the nervous system as a whole is engaged at the level of personality.

Robert Dilts ultimately introduces a sixth level, called *purpose*. It is a response to the existential questions of "Why?" and "For whom?" Because the individual is part of a greater system, the sense of his/her life should be something external to himself/herself. The level of purpose is this responsible for links between the individual and a broader perspective beyond the self. At the physiological level, Dilts advances controversial theories about a collective nervous system creating fields of interaction between people. He invokes the highly enigmatic, even esoteric conception of Rupert Sheldrak, which postulates the existence of a so-called morphogenetic field. The morphogenetic field, in Sheldrak's opinion, is responsible for unexplained phenomena such as changes in the behaviour of a portion of a population causing a change in the remaining portion in the absence of the possibility of communication between them. Sheldrak's conception has never gained the acceptance of scientists.

Robert Dilts claims that the particular levels of the pyramid impact one another. However, the top-down impact is significantly stronger, meaning that, for example, a change in beliefs and values impacts behaviour, and partially environment, but has a lesser impact on identity and purpose. Questions can be adapted to each level, intended to influence a person to make changes. Typical questions for the individual levels are:

Environment—"Where?" "When?"
Behaviour—"What?"
Capabilities—"How?"
Beliefs and values—"Why?"
Identity—"Who?"
Purpose—"For whom?"

Owing to these questions, the Dilts Pyramid is presented as a coaching tool (Dilts, 2003).

In assessing this theoretical proposition, first and foremost, we should point out the highly hypothetical explanation given at the physiological level. While the lower levels of the pyramid have an obviously neurophysiological basis, the higher ones, particularly that of purpose, are purely the author's speculation. The pyramid itself is one manner of ordering a person's experiences, and can be used in coaching activities, particularly life-coaching. Whether the model is accurate from a theoretical perspective should be determined via empirical study. Unfortunately, databases of scientific papers do not contain any articles on the Dilts Pyramid. As in the case of metaprograms, the Dilts Pyramid is an untested alternative conception in relation to scientific psychology.

NLP—evidence-based consistent framework or colourful marketing packet?

Empirical analysis of the primary assumptions underlying NLP demonstrates that they have no basis in fact. Theories about the existence of systems of representation and the possibility to discover them through analysis of EMs have not found confirmation in psychological studies. Other elements constituting the content of NLP courses, such as presuppositions, metamodel, anchors, metaprograms, and the Dilts Pyramid are either adapted scientific conceptions, or—more frequently—an idea without confirmation in systematic scientific studies. The same is true of other methods not discussed here, such as timelines, the Milton model, or reframing. Each of them features a mix of scientific and subjective content, and at times the weird ideas of authors and practitioners. NLP thus does not constitute a cohesive theory in which fundamental and verified assumptions lead to practical tips for users. It is rather a group of diverse and interesting-looking set of tools based on the subjective experience of users and dressed up in marketing mumbo jumbo. From the perspective of the evidence-based approach, we are looking at a less-than-credible proposition. The question thus arises: where does the popularity of this phenomenon come from?

First, NLP employs language that is attractive for the reader and the participant of development programmes. The term "neurolinguistic programming" suggest a strong grounding in neurobiology and linguistics. From the scientific perspective, all human actions are based in the activity of the nervous system, and are thus *neuro*. In this respect, NLP is no different from learning a foreign language, dance, eating with chopsticks, or any other activity. Placing the word "neuro" in the name gives it the veneer of science. Apart from the vague, and at times non-scientific considerations of Dilts about the foundation of his pyramid (see the section on the Dilts Pyramid), the creators of NLP do not address the activity of the human nervous system. Things are slightly different with the term "linguistic." Indeed, NLP tools directly refer to language, and some of them (metamodel, presupposition) are associated with the science of linguistics or psycholinguistics. In turn, the word "programming" is clearly overused. NLP practitioners talk about programming the mind, which in the best case is simply exerting influence, frequently to dubious effect.

The second cause of the popularity of NLP is the promise of quick and spectacular effects. At NLP courses, and particularly in short films on the internet, we are told we will cope with negative emotions, achieve our goals, have better relations with people, get rid of our fears and other psychological problems, and generally attain a higher quality of life. The effectiveness of NLP techniques is promoted primarily by its practitioners and advocates. However, there is precious little by way of documented studies. An exception is one meta-analysis on the application of NLP techniques in psychotherapy (Zaharia et al., 2015).

The third cause is linked to the "discovery" of the secrets of the mind, hidden to the mere mortal human. For example, observing the movements of eyes, one can, according to Grinder and Bandler, understand whether a given situation is recalled or constructed. If constructed, this can mean that someone is lying. The promise of recognition of lies through observation of eyeball movements is exceptionally attractive.

All of these factors contribute to the popularity of NLP. However, from the scientific perspective, we are dealing with a well-promoted set of techniques which come in small part from science, and their remainder has either not been verified at all, or the results have come back negative.

NLP—a practical guide for users

As we wrote in the introduction, the essence of evidence-based practices is the search for not just any evidence, but the best available evidence. In this chapter we have presented the evidence available on both the fundamental theoretical assumptions of NLP and the effectiveness of methods applied, as well as those concerning particular techniques. The majority of available evidence from higher levels of the hierarchy indicates that both the assumptions of NLP and the methods and techniques applied within the approach find no confirmation in empirical science. In opposition to the evidence we have presented is evidence located lower in the hierarchy, that is, subjective experiences of particular practitioners of NLP. At the same time, qualitative analysis of NLP techniques indicates that there are more cohesive and tested methods we may draw on science for.

Managers and engineers seeking to retain an evidence-based approach in working with people, during training exercises, or as part of self-development initiatives should rather look to methods whose effectiveness has been confirmed in multiple studies. This will save them time, as well as provide them with greater effectiveness and coherence in their undertakings.

Some common characteristics of pseudoscientific approaches, which we consider NLP to be, allow practitioners to effectively identify them, and facilitate selecting the most effective. Pseudoscientific methods based on evidence low in the hierarchy or simply without any evidence in support most frequently:

1. promise rapid effects incongruent with those which are presently applied broadly;
2. are very frequently characterised as breakthrough methods in a given field of knowledge;
3. develop primarily outside of mainstream research in a given field, because science rather quickly establishes their usefulness and rejects them as unpromising;
4. explain the reluctance of scientists to engage them in terms of threat, conservatism, and the absence of sufficient competence to assess "innovative" methods;
5. as opposed to science, frequently have little or no roots in the past, which is visible in the lack of bibliographies in papers by their proponents;
6. create an internal jargon that is frequently incomprehensible for outsiders;
7. create (usually at a cost) systems for education and certification, and their methods are frequently subject to copyrights, patents, and trademarks.

The greater the number of these traits characterise the present approach on the market to trainings, the higher the probability that we are dealing with pseudoscience packed in a colourful, shiny box whose contents will not live up to the promises of its marketing.

References

Baddeley, M. (1989). Neurolinguistic programming: The academic verdict so far. *Australian Journal of Clinical Hypnotherapy and Hypnosis, 10*, 73–81.

Bandler, R., & Grinder, J. (1979). *Frogs into princes*, Moab, UT: Real People Press.

Bateson, G. (1979). *Mind and nature: A necessary unity (advances in systems theory, complexity, and the human sciences)*. Cresskill, NJ: Hampton Press.

Baumeister, R. F., Bratslavsky, E., Finkenauer, C., & Vohs, K. D. (2001). Bad is stronger than good. *Review of General Psychology, 5*, 323–370.

Beyerstein, B. L. (2001). Fringe psychotherapies: The public at risk. *The Scientific Review of Alternative Medicine, 5*, 70–79.

Brown, N. (2004). What makes a good educator? The relevance of meta programmes. *Assessment & Evaluation in Higher Education, 29*, 515–533.

Cialdini, R. B. (1980). Full-cycle social psychology. *Applied Social Psychology Annual, 1*, 21–47.

Devilly, G. J. (2005). Power therapies and possible threats to the science of psychology and psychiatry. *Australian and New Zealand Journal of Psychiatry, 39*, 437–445.

Dilts, R. (1983). *Roots of neurolinguistic programming*. Capitola, CA: Meta Publications.

Dilts, R. (1998). *Modeling with NLP*. Capitola, CA: Meta Publications.

Dilts, R. (2003). *From coach to Awakener.* Capitola, CA Meta Publications.

Dorn, F. J., Brunson, B. I., Bradfor, I., & Atwater, M. (1983). Assessment of primary representational systems with neurolinguistic programming: Examination of preliminary literature. *American Mental Health Counselors Association Journal, 5,* 161–168.

Einspruch, E. L., & Forman, B. D. (1988). Neurolinguistic Programming in the treatment of phobias. *Psychotherapy in Private Practice, 6,* 91–100.

Grinder, J., & Bandler, R. (1976). *The structure of magic II.* Palo Alto, CA: Science and Behavior Books.

Grinder, J., & Bandler, R. (1981). *Transe-formations: Neuro-linguistic Programming^{TM} and the structure of hypnosis.* Moab, UT: Real People Press.

Heap, M. (1988). Neurolinguistic programming: An interim verdict. In M. Heap (Ed.), *Hypnosis: Current clinical, experimental and forensic practices* (pp. 268–280). London, UK: Croom Helm.

Heap, M. (2008). The validity of some early claims of neurolinguistic programming. *Skeptical Intelligencer, 11,* 1–8.

Hoag, J. D. (2018). NLP meta programs [Blog post]. Retrieved from http://nlpls.com/articles/metaPrograms.php

Horn, L. R. (1996). Presupposition and implicature. In S. Lappin (Ed.), *The handbook of contemporary semantic theory* (pp. 299–320). Blackwell Handbooks in Linguistics. Oxford: Blackwell.

Korzybski, A. (1933). A non-Aristotelian system and its necessity for rigour in mathematics and physics. In: *Science and Sanity: An introduction to non-Aristotelian systems and general semantics* (pp. 747–761). Lakeville, CT: International Non-aristotelian Library Publishing Co. www.bibsonomy.org/bibtex/1d8845e1a33fdcf1c6f18b1cb2d37c0c1/voj

Krugman, M., Kirsch, I., Wickless, C., Milling, L., Golicz, H., & Toth, A. (1985). Neuro-Linguistic Programming treatment for anxiety: Magic or myth? *Journal of Consulting and Clinical Psychology, 53,* 526–530.

Lankton, S. (1980). *Practical magic,* Cupertino, CA: Meta Publications.

Methods of neuro-linguistic programming. (n.d.). In Wikipedia. Retrieved October 20, 2018, from https://en.wikipedia.org/wiki/Methods_of_neuro-linguistic_programming

Moore, W. J., Manning, S. A., & Smith, W. I. (1978). *Conditioning and instrumental learning.* New York, NY: McGraw-Hill Book Company, pp. 52–61.

Newbrook, M. (2008). Linguistic aspects of "Neurolinguistic programming". *Skeptical Intelligencer, 11,* 27–29.

Ng, C., & Benedetto, U. (2016). Evidence hierarchy. In G. Biondi-Zoccai & G. Biondi-Zoccai (Eds.), *Umbrella reviews: Evidence synthesis with overviews of reviews and meta-epidemiologic studies* (pp. 11–19). Cham, Switzerland: Springer International Publishing.

Norcross, J. C., Koocher, G. P., & Garofalo, A. (2006). Discredited psychological treatments and tests: A Delphi Poll. *Professional Psychology: Research and Practice Copyright, 37,* 515–522.

O'Connor, J., & Seymour, J. (1993). *Introducing neuro-linguistic programming: Psychological skills for understanding and influencing people.* London, UK: Thorsons.

Page, S., & Meerabeau, L. (2004). Hierarchies of evidence and hierarchies of education: Reflections on a multiprofessional education initiative. *Learning In Health & Social Care, 3*(3), 118–128. doi:10.1111/j.1473–6861.2004.00070.x

Petrisor, B.A, & Bhandari, M. (2007). The hierarchy of evidence: Levels and grades of recommendation. *Indian Journal of Orthopaedic, 41*(1), 11–15.

Roderique-Davies, G. (2009). Neuro-linguistic programming: Cargo cult psychology? *Journal of Applied Research in Higher Education, 1,* 57–63.

Sharpley, C. F. (1984). Predicate matching in NLP: A review of research on the preferred representational system. *Journal of Counseling Psychology, 31,* 238–248.

Sharpley, C. F. (1987). Research findings on neurolinguistic programming: Nonsupportive data or untestable theory? *Journal of Counseling Psychology, 34,* 103–107.

Sturt, J., Ali, S., Robertson, W., Metcalfe, D., Grove, A., Bourne, C., & Bridle, C. (2012). Neurolinguistic programming: A systematic review of the effects on health outcomes. *The British Journal of General Practice, 62*(604), e757–e764. doi:10.3399/bjgp12X658287

Swets, J. A., & Bjork, R. A. (1990). Enhancing human performance: An evaluation of "New Age" techniques considered by the U.S. army. *Psychological Science, 1,* 85–86.

Vaknin, S. (2010). *The Big Book of NLP, Expanded: 350+ techniques, patterns & strategies of neuro linguistic programming.* Inner Patch Publishing.

Von Bergen, C. W., Soper, B., Rosenthal, L. V., & Wilkinson, L. V. (1997). Selected alternative training techniques in HRD. *Human Resource Development Quarterly, 8,* 281–294.

Witkowski, T. (2010). Thirty-five years of research on neuro-linguistic programming. NLP research data base. State of the art or pseudoscientific decoration? *Polish Psychological Bulletin, 41*(2), 58–66.

Witkowski, T. (2012). A review of research findings on neuro-linguistic programming. *The Scientific Review of Mental Health Practice, 9*(1), 29–40.

Zaharia, C., Reiner, M., & Schütz, P. (2015). Evidence-based neuro linguistic psychotherapy: A meta-analysis. *Psychiatria Danubina, 27*(4), 355–363.

chapter five

The evolution of Cultural Intelligence (CQ) and its impact on individuals and organisations

Cláudia Sofia Fanha Moura and Carolina Machado
University of Minho

Contents

Introduction

Cultural Intelligence (CQ) is a strategic and competitive advantage in organisations in that it makes individuals more effective when they are in different multicultural contexts. This chapter will explain the evolution of CQ, starting with a definition of intelligence, Social Intelligence and Emotional Intelligence. Subsequently, and in a more detailed way, we will talk about CQ, its profiles and its main characteristics, ending with an approach to recent developments. It turns out that a culturally intelligent individual is more effective and more successful in transcultural settings, as well as better performance and adaptability.

"In a world where crossing boundaries is routine, CQ becomes a vitally important aptitude and skill, and not just for international bankers and borrowers" (Earley & Mosakowski, 2004, p. 1). The culture of a given country significantly influences the present environment in organisations, particularly in economic, social, political, environmental and religious terms (Emmerling & Boyatzis, 2012). In an organisation that highlights the human resources diversity, it is necessary to know how to manage, in the best way, different individuals. A new concept emerges to represent cross-cultural competencies and the sensitivity needed by individuals to adapt to a new culture. This concept is called *Cultural Intelligence* (Rego & Cunha, 2009).

CQ can then be defined as "(…) an individual capacity, allowing one to more effectively interact with a variety of cultural settings; thus representing an advancement that can help to better situate individuals for a variety of inter-cultural interactions" (MacNab & Worthley, 2012, p. 62). As such, the study of CQ refers to the treatment and analysis of behaviours, seen as intelligent by individuals, in a different cultural context. It is believed that CQ can be the solution of the twenty-first century to manage effectively in a multicultural context (Jyoti & Kour, 2017) and to understand why some individuals perform better than others when inserted in a different cultural context (Dogra & Dixit, 2017; Hofstede, 2011).

The first section of this chapter concerns the explanation of intelligence per se and the definition of Social Intelligence. Next, Emotional Intelligence will be explained, namely the difference between reason and emotion, the definition of the concept and the main results underlying the use of Emotional Intelligence. Later, CQ will be explained in a more concrete way, beginning with the relationship between social, Emotional and Cultural Intelligence, through the definition of the concept of CQ, main characteristics and explanation of CQ profiles. Finally, a final topic is made about recent developments.

Literature review

Intelligence per se (IQ) and Social Intelligence (SI)

Intelligence (IQ)—The concept

The concept of intelligence (I or IQ), also called general mental ability, g-factor or even general cognitive ability, is a concept that has been tried to be understood for many years. In 1944, David Wechsler defined intelligence as follows: "intelligence is the aggregate or global capacity of the individual to act purposefully, to think rationally, and to deal effectively with his environment" (Wechsler, 1958, p. 3). This is one of the most commonly used definitions because it is the one that looks at intelligence as broadly as possible (Salovey & Mayer, 1990). A recent definition of intelligence could be "an individual's ability to learn accurately and quickly a task, a subject or a skill, under optimal instructional conditions. Less time and greater accuracy indicate greater general mental capacity" (Salgado, Moscoso, & Lado, 2006, p. 113).

Considering Spearman's unifactorial model, or "g", a person is more intelligent the greater the amount of his "g" factor or general factor of intelligence (Woyciekoski & Hutz, 2009). To Spearman, the most important and essential measure to measure intelligence is the "g" factor, which is defined by the ability to have an abstract thought and by the aptitude of thinking (Sternberg & Wagner, 1992). It is important to highlight that intelligence by itself is a vast set of abilities. On the other hand, as models of intelligence they look to explain the causes and relationships existing between the diverse mental abilities, representing a more restricted field than intelligence. All mental abilities are considered as intelligence; however, they may or may not be related to each other. As an example, Emotional Intelligence is referred to as one of the existing types of intelligence, but it is not necessarily related to other types of intelligence (Salovey & Mayer, 1990).

Social Intelligence (SI)

The first proposed definition of SI appeared in 1920 and understood it as "the ability to understand men and women, boys and girls—to act wisely in human relations" (E. L. Thorndike, 1920, p. 227). Later, in a research developed by the same author together with S. Stein, they suggested that SI was "the ability to understand and manage people" (Thorndike & Stein, 1937, p. 275). In a more simplistic way, SI is then the ability of a person, namely the leader, in an organisational context, to be able to understand their own states of mind, motives and behaviours, as well as those of those around them in order to act in the best way, based on the information they have available. The fact that the concept includes the ability to understand and manage their own behaviours and attitudes, through their intellectual or social skills, leads to an extension to their own understanding (Salovey & Mayer, 1990).

However, literature is never a sea of roses for whatever the theme is, and as such there is also a more negative connotation associated with the concept. As stated by several authors, there is a manipulative connotation in SI because it is understood that the ability to perceive others may lead to a person being able to change other people's answers voluntarily, but for the sake of their own interests (Bureau of Personnel Administration, 1930). Moreover, it is believed that the manipulative character is also due to the omission of the emotions of the one's own and the other elements Dienstbier, 1984; Hoffman, 1984).

The fact is that this is a construct where the barriers between it and other types of intelligence are easily transposed being difficult to disassociate it, for example, from mechanical and abstract intelligence. For these and other reasons, for some years the definition and research on the subject were somewhat stagnant (Salovey & Mayer, 1990). However, in the eighties, Gardner based on Thorndike's studies formalised the theory of multiple intelligences consisting of seven types of intelligence, one of which was SI. Each type of intelligence consisted of two dimensions: the interpersonal and the intrapersonal. The interpersonal related to the ability to understand the feelings and desires of others and to act on the basis of this information, and the intrapersonal that involves the aptitude to understand itself and to be able to carry out a self-evaluation (Salovey & Mayer, 1990; Rego, Cabral-Cardoso, Cunha, M., & Cunha, R., 2007).

Emotional Intelligence (EI)

Reason and emotion

For many years, attention was given exclusively to reason and the emotions of the human being were ignored. However, the relationship between emotions, personal or family life and work is increasingly perceived. This trio is constituted by complex interpersonal relationships that overflow easily from the extraorganisational to the intraorganisational environment and vice versa (Rego et al., 2007). Duck (1993) was able to explain this phenomenon in an exemplary way saying that

> For decades, managers and workers have been told to check their feelings at the door. And that's a big mistake. It's one thing to say that behavior is more accessible to managers than feelings are; it's another thing altogether to say that feelings have no place at work. (…) Companies that want their workers to contribute with their heads and hearts have to accept that emotions are essential to the new management style. The old management paradigm said that at work people are only permitted to feel emotions that are easily controllable, emotions that can be categorized as "positive." The new management paradigm says that managing people is managing feelings. The issue isn't whether or not people have "negative" emotions; it's how they deal with them. (Duck, 1993, p. 113)

It should be noted that the literature on this subject is relatively recent, referring the first studies published to 1993 with Stephen Fineman's book *Emotion in Organisations,* to 1998 with Goleman writing the well-known book *Working with Emotional Intelligence* and only in 2001 was launched the journal *Emotion* (Rego et al., 2007). However, there is already a recognition that reason and emotion are not antagonistic concepts, on contrary, they complement each other. As referred by the first authors to give a definition of EI, emotions can serve as a source of information to the others. To these authors the information provided by the emotions can be adaptive and the relationship between emotion and thought is not necessarily antagonistic (Salovey, Mayer, Goldman, Turvey, & Palfai, 1995).

Emotional Intelligence (EI)—The definition

Before going to the exploration of the concept, it is important to distinguish emotions from humours. On the one hand, emotions are motivational forces that manifest before something or someone. They are feelings much more intense than humours and usually do not manifest for long periods of time (only few minutes). Emotions arise in response to internal or external situations and usually appear when someone says they are happy, sad or angry. On the other hand, humours usually occur less intensely but for longer periods of time (hours). In addition, there is no event, situation or specific person to cause any humour; usually, the individual cannot explain the why of his good or bad humour. It should be noted that emotions can become humours when the individual can not identify what has caused such emotion and humours can generate more emotional responses in relation to a situation or person (Salovey & Mayer, 1990; Robbins & Judge, 2013).

It was in the 1990s that the first authors appeared to formalise the concept of EI. For Salovey and Mayer, EI is a sub-dimension of SI (concept initially defined by Thorndike) and can be defined as "the ability to monitor one's own and other's feelings and emotions, to discriminate among them and to use this information to guide one's thinking and actions" (Salovey & Mayer, 1990, p. 190). In other words, it involves "the competence to perceive and express emotions, to understand and use them, and to manage them in oneself and in others" (Rego et al., 2007, p. 134). According to the same authors, EI comprises four competences. The first is the ability to perceive, express and evaluate emotions; the second is the ability to accelerate cognitive activities through the range and appearance of feelings; the third comprises the ability to analyse and evaluate the emotional messages transmitted, as well as to use this emotional knowledge and, finally, the ability to manage emotions in order to improve and develop one's own and others' intellectual and emotional well-being (Rego et al., 2007).

EI brings new contributions to the literature at various levels. Using EI makes the intelligence dimension a broader field where it is possible to find issues related not only to reasoning and thinking but also to people's feelings and emotions (Woyciekoski & Hutz, 2009). It is believed that the reconciliation of reason and emotion is a competitive advantage especially in an organisational context because it allows intelligent reasoning about emotions and at the same time uses emotions to help reason. In this way, "the emotion makes the thought more intelligent, and the intelligence allows to think and to use in a more accurate way the emotions" (Rego et al., 2007, p. 134).

It is easy to understand why there are authors who consider EI as a sub-dimension of SI. It should be noted that human beings are made up of many social needs and that the absence of realising these needs can lead to a restriction in the ability of humans to adapt to the social context. Individuals need social skills; however, they also need emotional skills so that their behaviour and adaptation to society is facilitated (Sternberg, 1997). Communication with other individuals and socialisation is facilitated by the use of emotions intelligently as they provide information about people's intentions and conceptions (Lopes et al., 2004).

However, with the release of Daniel Goleman's book, *Emotional Intelligence*, the concept gained wider scope and a new definition where terms relating to the individual's personality were denoted. EI thus becomes the ability to recognise and manage our emotions and those of others, in order to make our thinking smarter (Goleman, 1999)—namely a subset related to the perception of humours and emotions, known as *Theory of Mind* (ToM) (Baron-Cohen, Leslie, & Frith, 1985). ToM is a "mechanism which underlies a crucial aspect of social skills, namely being able to conceive of mental states: that is, knowing that other people know, want, feel, or believe things" (Baron-Cohen et al., 1985, p. 38). Still following Goleman's research, it can be said that concepts such as empathy, social dexterity, impulse control, persistence, self-awareness, self-motivation and enthusiasm are characteristic of EI (Goleman, 1998).

Main results of using EI

The use of EI in an organisational context generates results for the individual, for the group and even for the organisation. Individuals who develop their EI can express and perceive their own emotions as well as those of others by using them to perform in the best possible way and to promote the best behaviours in others as well as in oneself (Salovey & Mayer, 1990). The success or failure of social relationships between individuals is due in large part to the use of EI. In addition, it is believed that it is not enough to have a good IQ (coefficient of intelligence) but especially a good use of EI and that it is responsible for the improvements identified in the leaders' performance (Goleman, 1998). Leaders with emotional competencies tend to be more effective at achieving organisational goals and usually have more committed, more satisfied with work and professionally accomplished followers (Rego et al., 2015).

A relationship of dependence between IQ and EI was observed. There is a considerable variation in the individuals' performance when IQ and

EI are used together than when only IQ is applied. IQ is necessary, and even requires a minimum level for a difference in performance when applied to EI. However, the effects of EI use only occur in the presence of IQ (Rego et al., 2007).

Some results related to the groups were also identified. Groups where behaviours activated by EI are identified tend to have high performances, associated to the fact that they understand the emotions and humours of each individual belonging to the group. Individuals in an emotionally intelligent group are self-aware, seek feedback on their actions and when the functioning of the group is being breached there is a confrontation where brainstorming is allowed to exchange ideas (Druskat & Wolff, 2001).

Also for the organisational culture and for the organisation itself are improvements in results. A culture that is bathed by emotionally intelligent individuals has as characteristic to promote the identification of the members with the values of the organisation, that results in a greater commitment and motivation on the part of the same ones. In the same way, EI promotes organisational effectiveness and facilitates the balance of organisational culture, that is, the alignment between the interests of the organisation and those of each individual belonging to it (Ugoani, 2015).

Cultural Intelligence (CQ)

Existing relationships between the triad

It is important to explain the relationship between the triad—CQ, SI and EI. There are points of contact between this triad, however, it should be noted that they are different concepts. As mentioned earlier, SI is the ability of an individual to understand the others' attitudes and behaviours as well as his/her own in order to create relationships with the people around them (Thorndike & Stein, 1937; Cantor & Kihlstrom, 1985). By its turn, EI is the ability to recognise and manage our emotions and those of others, in order to make our thinking smarter (Goleman, 1999). It is said that a combination of SI and EI generates more effective individuals; however, this only occurs in their own cultures. This is the main characteristic that differentiates these types of intelligence; an "alien" who only has social and emotional skills cannot be effective in a new culture. But if he/she also has CQ he/she can improve his/her skills in an environment full of transculturality. Both SI and EI are loaded with culture, for example, the ability to respond appropriately to other individuals, or the ability to generate empathy in the other through the emotions are culturally influenced (Brislin, Worthley, & Macnab, 2006).

Earley and Mosakowski (2004) found that CQ begins where EI ends. According to Witzel, Rohde, and Brushart (2005) a person with a high EI can be completely unable to understand, in different cultural contexts, different signs and movements. An emotionally intelligent individual understands each person as a different being with different behaviours and emotions. It promotes empathy among individuals and attempts to manage behaviours in order to achieve consensus among all, while a culturally intelligent individual manages to generate behaviours that all people identify with, even if they are not universal (Earley & Mosakowski, 2004). However, it turns out that both types of intelligence share something, such as the "propensity to suspend judgment—to think before acting" (Goleman, 1998, p. 90).

In this way, CQ is considered a higher level of SI by allowing an individual to be socially effective, but in different cultural environments (Brislin et al., 2006), as well as an extension or enhancement of EI (Earley & Mosakowski, 2004).

Cultural Intelligence—Definition and characteristics

One of the main difficulties individuals encounter when changing their country or even their organisation is to adapt to their culture. Organisations are spaces rich in diversity where there are not only different people as different departments and where we can find subcultures within the organisational culture. A new employee usually takes the early days trying to adapt to the new reality. Some take some time, however, others have a natural ability to understand gestures, attitudes and behaviours to which they are not accustomed. This capacity is called Cultural Intelligence (CQ) (Rego & Cunha, 2009).

CQ is a "multidimensional construct targeted at situations involving cross-cultural interactions arising from differences in race, ethnicity and nationality" (Ang et al., 2007, p. 336). In other words, "cultural intelligence (also known as "CQ") refers to an individual capacity, allowing one to more effectively interact with a variety of cultural settings, thus representing an advancement that can help to better situate individuals for a variety of inter-cultural interactions" (MacNab & Worthley, 2012, p. 62). A culturally intelligent individual is able to easily adapt to a new cultural context, is able to understand interactions and behaviours typical of other cultures, has a capacity to adapt his own behaviours according to the context, effectively perceives gestures and intercultural behaviours as if they were the same who exist in their own culture and have the ability to act and behave as if they are all familiar (Brewster, Houldsworth, Sparrow, & Vernon, 2016). One of the main difficulties when a person faces a different culture is to perceive what is, and what is not acceptable to them. This difficulty is easily overcome by an individual who possesses CQ, as through the observation of the behaviours, reactions and attitudes of other people, this individual draws his elations and adapts to the context (Rego & Cunha, 2009).

Triandis' suspension judgements (2006) and the acceptance of Brislin confusion (2006) are two fundamental characteristics of CQ that differentiate it from social and Emotional Intelligence. Suspension of judgements is related to an individual's ability to delay making judgements about something or someone from another culture until they have the information they feel is sufficient to proceed with that judgement. Individuals are very different from each other, especially when the cultural context changes and as such, it is necessary to gain a great amount of information about them so that the judgement made does not end up being skewed. A culturally intelligent individual looks not only at the evidence provided by culture but also at the idiosyncrasies of each person that are as or more important than the culture in which he is inserted (Triandis, 2006). After analysing the individual characteristics of the people, it is also important to take into account the situation where the interaction took place, since it also provides important information to make a correct judgement (Chatman & Barsade, 1995).

Associated with the ability to not make immediate judgements, the culturally intelligent individual must also accept the confusion and the fact that there may be misunderstandings between people because they do not know everything. It is expected for this individual the existence of disagreements by the cultural differences existing between the people. So, the same opts to adopt the *"zen"* mode thus reducing the feelings of stress and mistrust in the interaction. The acceptance of confusion is thus defined as the "willingness to accept not knowing (…) that will then allow the sojourner to better evaluate the situation leading to eventual, and more accurate, understanding" (Brislin et al., 2006, p. 49).

It should be noted that although many of the competencies of CQ depend on the characteristics of each individual, it is always possible to improve them. It is enough that for this the person has the will to learn, is motivated and wants to adapt to the new context (Rego & Cunha, 2009). However, it involves an extra effort to train its head, body and heart (Earley & Ang, 2003), as well as learn to suspend jugdments and accept confusion in order to promote desired behaviours and eliminate those that generate conflicting situations (Paige & Martin, 1996). Positive attitudes, experience and practice in multicultural contexts are other characteristics that positively influence the learning of CQ (Brislin et al., 2006).

Cultural Intelligence sources

The three sources of CQ are the head (cognitive), the body (physical) and the heart (emotional/motivational). The cognitive dimension is the one that is usually most evident when a new employee joins the organisation or the new arrival of one person to another country. Individuals with high levels in the cognitive component can find the differences and similarities between different cultures (Brislin et al., 2006). Usually to these individuals are offered learning strategies in the form of habits, values, taboos and customs of the new culture that aim to sensitise them to the particularities of that culture (Rego & Cunha, 2009).

But this awareness does not guarantee that the individual acts as expected and therefore it is also necessary to promote the actions necessary to show that there is already an adaptation to the context. These actions can be expressed, for example, by acquiring the ways and habits of the people of that culture who, in turn, promote greater openness and confidence of the native people towards the foreigner. To these actions correspond the dimension of the physical body of CQ (Earley & Mosakowski, 2004). High levels in this component allow the person to adapt and adopt expressions, gestures and words from another culture due to their huge repertoire of behaviours (Gudykunst, Matsumoto, Ting-Toomey, & Heyman, 1996).

Last but not least, the dimension of the heart. This dimension emphasises the fact that the individual has many obstacles in his/her way to adapt and he/she has two solutions: either he/she discourages and gives up in making more efforts to adapt because he/she does not find the results that he/she wants; or he/she uses these obstacles to increase his/her motivation and to overcome them more effectively increasing his/her confidence, becoming independent of the results that may arise (Earley & Mosakowski, 2004). The heart directs the individual's attention to his/her self-efficacy in a multicultural context (Bandura, 2001).

Recently, a new dimension has been added to the construct called cognition. Cognition, or metacognitive dimension, provides information about individuals' knowledge about different cultures (Brewster et al., 2016), as well as the processes they use to obtain and understand this knowledge (Ang et al., 2007). This component allows the individual to know other cultures and adjust his/her behaviours before, during and after contact with other people (Brislin et al., 2006).

It should be noted that an individual does not have to have the three dimensions on the same level. There may be one that stands out but that compensates for the lower level of the other dimensions (Earley & Mosakowski, 2004). It is concluded that individuals with high CQ, and consequently high values in their sources, are better able to communicate with other cultures, showing more efficacy and success in outbound trips, as well as a better capacity to deal with change (Brewster et al., 2016).

Cultural Intelligence profiles

As explained above, the level assigned to each dimension does not have to be the same. As such, there are seven possible combinations depending on these levels, giving origin to the profiles of CQ. Among them are the provincial, the analyst, the natural, the ambassador, the mimic, the naive and the chameleon (Rego & Cunha, 2009). It should be noted that these types of profiles are used to characterise managers or leaders when they are in another culture and that only three dimensions are used to characterise the CQ profiles because the fourth one (cognition) is still recent in the literature, not existing in studies that prove its influence on the profiles.

It is understood as provincial the individuals who have low levels in the three dimensions. The provincial is very effective when the cultural picture belongs to him/her, but shows many gaps when he/she is in another culture because he/she is not able to understand and act appropriately with people of other cultures. In a different cultural framework, he/she cannot build trust in other people and ends up generating conflicts and mistrust in others.

The analyst, also known as the "cognitive animal" (Rego & Cunha, 2009), is the one who easily perceives that he/she is in another environment and that analyses the specific characteristics of this culture to then formulate learning strategies for newcomers. The analyst is more flexible allowing to change the strategy when he/she finds resistance to the existent one at a given time. In this way, instead of forcing individuals to fulfill a certain strategy to achieve the objectives, he/she adapts it to achieve the same objective differently (Earley & Mosakowski, 2004; Rego & Cunha, 2009).

Named natural or intuitive, this profile is known to have individual who depend on their intuition to define the strategies they will use through the capture of signals existing in the environment. However, there is an inherent problem to this profile. Intuitives tend to fail when there is a very ambiguous environment because they are not prepared to create new learning strategies to interact with the natives, leading to disorientation on the part of the individual with a natural profile.

Known for having political characteristics is the ambassador. This type of profile is the most frequent of all profiles and has as main characteristic the trust of the individual. Although being a profile in which the knowledge of the surroundings and the culture is not great, it excels to have a lot of confidence when adapting, which gives it evident characteristics of emotional and Cultural Intelligence. Most of the knowledge of these individuals comes from the observation of other cases that have succeeded in similar situations; however, despite their confidence, they also have the humility to know that they have a lot of ignorance about the characteristics of the new culture where it is inserted (Earley & Mosakowski, 2004; Rego & Cunha, 2009).

Named by mimic, and as the name implies, it is the one that tries to adopt a style of interaction and speech similar to that of the culture where it is inserted. Being able to behave in a similar way to the native people makes communication and trust easier. This profile is also characterised by the excessive control of their behaviours and actions and the rapid capture of the signals given by the environment, providing a natural adjustment to the individual of this profile.

Naive or *"motivated animal"* (Rego & Cunha, 2009) is the name of another type of profile characterised by an enormous desire to become a part of the new culture. However, it has difficulty in constantly making an effort to change its behaviour and to understand other cultures, which makes adjustment difficult.

And, last but not the least, the chameleon. If at the beginning of this section, the provincial was known with low levels in the three components of CQ, the chameleon is exactly the opposite. This type of profile is the rarest of all existing and usually defines individuals who characterise themselves as *"citizens of the world"* (Rego & Cunha, 2009). These individuals can easily be confused with a native of the culture by their ability to see both internally and externally, which gives them very good managerial qualities. They are known for achieving high results and for adapting their behaviours and attitudes according to the cultural framework in which they are (Earley & Mosakowski, 2004; Rego & Cunha, 2009).

According to the aforementioned study by Earley and Mosakowski (2004), a manager can have characteristics of several profiles at the same time, forming a hybrid. He also suggests that the most common hybrid is the one that combines the analyst and the ambassador.

Latest developments

In recent years, there has been an increase in researches related to CQ. In a study by Ang and his co-authors conducted in 2007, we sought to understand the relation between the four competences of CQ, namely metacognitive, cognitive, behavioural and emotional and cultural adaptation, cultural decision-making and judgement, and the performance of tasks in multicultural contexts. These three results correspond, respectively, to affective, cognitive and behavioural effectiveness. The relationship between the variables was empirically proven, and the influence of CQ (as a set of dimensions) on the different results was notorious. It was also possible to note that there are some practical implications "especially for selecting, training and developing a culturally intelligent workforce" (Ang et al., 2007, p. 365).

In more recent studies, it was also possible to observe the existence of a relationship between CQ and leadership. For Rockstuhl, Seiler, Ang, Dyne, and Annen (2011), the effectiveness of a leadership that occurs across borders is improved with greater level of CQ of the leader. This high degree of CQ enables the leader to demonstrate greater performance when in multicultural teams than in uniform teams. Consequently, multicultural teams also demonstrate a higher performance than uniform teams (Groves & Feyerherm, 2011). Similarly, when it comes to negotiations, it is also found that the performance of highly culturally minded and interest-oriented negotiators was superior to negotiators with low levels of CQ (Groves, Feyerherm, & Gu, 2014).

In 2017, new contributions brought, again, the four dimensions of CQ to the researches. In the article by Dogra and Dixit (2017), it was possible to find a positive and statistically strong relationship between the variables CQ and innovation. It was concluded that high levels of CQ generated better performing teams and had a strong potential to explain why there are individuals who learn and perform better in the international context than others. There is also another article that, in a way, confirms the results achieved by these authors. When CQ and an individual's ability to adapt to multicultural contexts are highlighted, there is a strong and positive relationship between the variables. In this way, the "cross-cultural adaptability" (Jyoti & Kour, 2017, p. 308) of individuals was found to be considerably better when individuals were considered culturally intelligent (Jyoti & Kour, 2017).

The same authors continued their research and published another article of great relevance to this work. Among the obtained results, the influence of EI on CQ is highlighted because emotionally intelligent individuals can understand their own emotions as well as those of others, which contributes to also perceive them in another cultural context. There is also a positive relationship between Social and Cultural Intelligence as socially intelligent individuals have a very accurate sensitivity to difficult situations which makes them more effective in multicultural environments. Another result that the authors came up with was that "cross-cultural adaptability" (Jyoti & Sumeet, 2017, p. 781) actually mediates the relationship between CQ and worker performance, which strengthens previous research (Jyoti & Sumeet, 2017).

Conclusion

Social and Emotional Intelligence are very relevant constructs and of great contribution to the effectiveness of the management in an organisational context (Emmerling & Boyatzis, 2012). SI allows the construction of relationships with other individuals, as well as the understanding of the behaviours and attitudes of others and of the self (Thorndike & Stein, 1937). EI is related to the knowledge and management of the emotions of the individual and others, so as to use them in the most intelligent way, in a social context (Goleman, 1999).

Despite their relevant contributions, the results are only found in a national culture. When going out into a different culture, it is necessary more than understanding people's behaviour, attitudes and emotions. And it is in this sense that CQ emerges as an enhancement of the aforementioned intelligences (Earley & Mosakowski, 2004). CQ provides an ability to understand the attitudes, gestures and behaviours of people from other cultures in a way that facilitates interactions between them

(Jyoti & Kour, 2017). Two of the key distinguishing characteristics of CQ are that individuals have the ability to suspend judgement until they have enough information to draw conclusions with as few biases as possible (Triandis, 2006) and the acceptance of the confusion related to the ability of the culturally intelligent individual to understand that there may always be misunderstandings, especially when talking about interactions between people of different cultures(Brislin et al., 2006).

It is believed that an individual with various types of intelligence is a more effective and successful individual in cross-cultural contexts and as such, CQ comprises a competitive and strategic advantage for organisations (Yitmen, 2013). In addition, there is improvement in leadership, multicultural groups and individuals. There is an increase in people's performance and a greater ability to adapt to new cultures (Dogra & Dixit, 2017). These are some of the possible results to prove when individuals are culturally intelligent; however, more research needs to be done in order to reinforce empirical research on the subject.

References

Ang, S., Dyne, L. V., Koh, C., Ng, K. Y., Templer, K. J., Tay, C., & Chandrasekar, N. A. (2007). Cultural intelligence: Its measurement and effects on cultural judgment and decision making, cultural adaptation and task performance. *Management and Organisation Review, 3*(3), 335–371.

Bandura, A. (2001). Social cognitive theory: An agentic perspective. *Annual Review of Psychology, 52,* 1–26.

Baron-Cohen, S., Leslie, A. M., & Frith, U. (1985). Does the autistic child have a "theory of mind"? *Cognition, 21*(1), 37–46.

Brewster, C., Houldsworth, E., Sparrow, P., & Vernon, G. (2016). *International human resource management* (4ᵃ ed.). London, UK: Chartered Institute of Personnel and Development.

Brislin, R., Worthley, R., & Macnab, B. (2006). Cultural intelligence: Understanding behaviors that serve people's goals. *Group & Organisation Management, 31*(1), 40–55.

Bureau of Personnel Administration. (1930). Partially standardized tests of social intelligence. *Public Personnel Studies, 8,* 73–79.

Cantor, N., & Kihlstrom, J. F. (1985). Social intelligence: The cognitive basis of personality. In P. Shaver (Ed.), *Review of personality and social psychology* (Vol. 6, pp. 15–33). Beverly Hills, CA: Sage.

Chatman, J. A., & Barsade, S. G. (1995). Personality, organisational culture, and cooperation: Evidence from a business simulation. *Administrative Science Quarterly, 40,* 423–443.

Dienstbier, R. A. (1984). The role of emotion in moral socialization. In C. E. Izard, J. Kagan, & R. B. Zajonc (Eds.), *Emotions, cognition, and behavior* (pp. 484–514). New York, NY: Cambridge University Press.

Dogra, A. S., & Dixit, V. (2017). Leveraging from cultural quotient: Linking cultural intelligence (CQ) with team performance (innovation). *Proceedings of ICRBS,* 682–687.

Druskat, V., & Wolff, S. (2001). Building emotional intelligence of groups. *Harvard Business Review, 79*(3), 81–90.

Duck, J. D. (1993). Managing change: The art of balancing. *Harvard Business Review, 71*(6), 109–118.

Earley, P. C., & Ang, S. (2003). *Cultural intelligence: Individual interactions across cultures.* Stanford, CA: Stanford Business Books.

Earley, P. C., & Mosakowski, E. (2004). Cultural Intelligence. *Harvard Business Review, 82*(10), 139–146.

Emmerling, R. J., & Boyatzis, R. E. (2012). Emotional and social intelligence competencies: Cross cultural implications. *Cross Cultural Management, 19*(1), 4–18.

Goleman, D. (1998). What makes a leader? *Harvard Business Review, 76*(6), 93–102.

Goleman, D. (1999). *Working with emotional intelligence.* London, UK: Bloomsburry.

Groves, K. S., & Feyerherm, A. E. (2011). Leader cultural intelligence in context: Testing the moderating effects of team cultural diversity on leader and team performance. *Group & Organisation Management, 36*(5), 535–566.

Groves, K. S., Feyerherm, A., & Gu, M. (2014). Examining cultural intelligence and cross-cultural negotiation effectiveness. *Journal of Management Education, 39*(2), 209–243.

Gudykunst, W. B., Matsumoto, Y., Ting-Toomey, S., & Heyman, S. (1996). The influence of cultural individualism-collectivism, self construals, and individual values on communication styles across cultures. *Human Communication Research, 22*(4), 510–543.

Hoffman, M. L. (1984). Interaction of affect and cognition in empathy. In C. E. Izard, J. Kagan, & R. B. Zajonc (Eds.), *Emotions, cognition, and behavior* (pp. 103–131). New York, NY: Cambridge University Press.

Hofstede, G. (2011). Dimensionalizing cultures: The Hofstede model in context. *Online Readings in Psychology and Culture, 2*(1), 1–26.

Jyoti, J., & Kour, S. (2017). Factors affecting cultural intelligence and its impact on job performance. *Personnel Review, 46*(4), 767–791.

Jyoti, J., & Sumeet, K. (2017). Cultural intelligence and job performance: An empirical investigation of moderating and mediating variables. *International Journal of Cross Cultural Management, 17*(3), 305–326.

Lopes, P. N., Brackett, M. A., Nezlek, J. B., Schutz, A., Sellin, I., & Salovey, P. (2004). Emotional intelligence and social interaction. *Personality and Social Psychology Bulletin, 30*(8), 1018–1034.

MacNab, B. R., & Worthley, R. (2012). Individual characteristics as predictors of cultural intelligence development: The relevance of self-efficacy. *Internacional Journal of Intercultural Relations, 36*, 62–71.

Paige, R. M., & Martin, J. N. (1996). Ethics in intercultural training. In D. Landis, & R. S. Bhagat (Eds.), *Handbook of intercultural training* (pp. 35–60). Thousand Oaks, CA: Sage.

Rego, A., Cabral-Cardoso, C., Cunha, M. P., & Cunha, R. C. (2007). *Manual de Comportamento Organizacional e Gestão* (6ª ed.). Lisboa: Editora RH.

Rego, A., & Cunha, M. P. (2009). *Manual da Gestão Transcultural de Recursos Humanos* (1ª ed.). Lisboa: RH Editora.

Rego, A., Cunha, M. P., Gomes, J. F., Cunha, R. C., Cabral-Cardoso, C., & Marques, C. A. (2015). *Manual de Gestão de Pessoas e do Capital Humano* (3ª ed.). Lisboa: Edições Sílabo.

Robbins, S. P., & Judge, T. A. (2013). *Organisational behavior* (15ª ed.). Upper Saddle River, NJ: Pearson Education.

Rockstuhl, T., Seiler, S., Ang, S., Dyne, L. V., & Annen, H. (2011). Beyond general intelligence (IQ) and emotional intelligence (EQ): The role of cultural intelligence (CQ) on cross-border leadership effectiveness in a globalized world. *Journal of Social Issues, 67*(4), 825–840.

Salgado, J. F., Moscoso, S., & Lado, M. (2006). Reclutamiento y selección. In J. Bonache & A. Cabrera (Eds.), *Dirección estratégica de personas* (pp. 101–137). Madrid: Financial Times/Prentice Hall.

Salovey, P., & Mayer, J. D. (1990). Emotional intelligence. *Imagination, Cognition and Personality, 9*(3), 185–211.

Salovey, P., Mayer, J. D., Goldman, S., Turvey, C., & Palfai, T. (1995). Emotional attention, clarity and repair: Exploring emotional intelligence using the trait meta-mood scale. In J. D. Pennebaker (Ed.), *Emotion, disclosure, and health* (pp. 125–154). Washington, DC: American Psychological Association.

Sternberg, R. J. (1997). Tacit knowledge and job sucess. In R. J. Sternberg (Ed.), *International handbook of selection & assessment* (pp. 201–213). London, UK: Wily.

Sternberg, R. J., & Wagner, R. K. (1992). Tacit knowledge: An unspoken key to managerial success. *Creativity and Innovation Management, 1*(1), 5–13.

Thorndike, E. L. (1920). Intelligence and its uses. *Harper's Magazine, 140,* 227–235.

Thorndike, R. L., & Stein, S. (1937). An evaluation of the attempts to measure social intelligence. *Psychological Bulletin, 34,* 275–284.

Triandis, H. (2006). Cultural intelligence in organisations. *Group & Organisation Management, 31*(1), 20–26.

Ugoani, J. N. (2015). Emotional inelligence and organisational culture equilibrium - A correlation analysis. *Journal of Advances in Social Science-Humanities, 1*(1), 36–47.

Wechsler, D. (1958). *The measurement and appraisal of adult intelligence.* Baltimore, MD: Williams & Wilkins.

Witzel, C., Rohde, C., & Brushart, T. M. (2005). Pathway sampling by regenerating peripheral axons. *Journal of Comparative Neurology, 485*(3), 183–190.

Woyciekoski, C., & Hutz, C. S. (2009). Inteligência emocional: teoria, pesquisa, medida, aplicações e controvérsias. *Psicologia: Reflexão e Crítica, 22*(1), 1–11.

Yitmen, I. (2013). Organisational cultural intelligence: A competitive capability for strategic alliances in the international construction industry. *Project Management Journal, 44*(44), 5–25.

Index

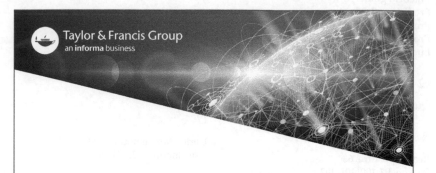

Milton Keynes UK
Ingram Content Group UK Ltd.
UKHW040051071024
449327UK00019B/467